ADVANCED TECHNOLOGIES IN BIODIESEL

ADVANCED TECHNOLOGIES IN BIODIESEL

INTRODUCTION TO PRINCIPLES AND EMERGING TRENDS

AMINUL ISLAM, YUN HIN TAUFIQ-YAP
AND ENG-SENG CHAN

MOMENTUM PRESS

MOMENTUM PRESS, LLC, NEW YORK

Advanced Technologies in Biodiesel: Introduction to Principles and Emerging Trends
Copyright © Momentum Press®, LLC, 2015.

First published by Momentum Press®, LLC
222 East 46th Street, New York, NY 10017
www.momentumpress.net

ISBN-13: 978-1-60650-502-1 (print)
ISBN-13: 978-1-60650-503-8 (e-book)

Momentum Press Thermal Science and Energy Engineering Collection

DOI: 10.5643/9781606505038

Cover and interior design by Exeter Premedia Services Private Ltd., Chennai, India

10 9 8 7 6 5 4 3 2 1

Printed in the United States of America

ABSTRACT

The important strategic issue of the 21st century states as the struggle for existence is the struggle for sustainable energy. In the last few years, the interest in renewable fuels has increased dramatically due to high demand of energy and the limitation of fossil fuel. Given the rapidly increasing demand for energy, which is projected to double by mid- 21st century, it is expected that biodiesels will become an important part of the global energy mix and make a significant contribution to meeting energy demand.

Through extensive research, many commercial enterprises have offered comprehensive, innovative, and state-of-the-art technologies to produce high-quality biodiesel consistently at a competitive price via transesterification process. Therefore, this book gives a critical review on the recent emerged process intensification technologies for biodiesel production as well as the various methods for assessing biodiesel fuel quality and/or monitoring the transesterification reaction with advantages and drawbacks, and offers suggestions on selection of appropriate methods, which could provide a thrilling adventure ahead of all interested scientists.

The adequate and up-to-date information provided in this book should be of interest for biochemical engineers, academics, post graduate and graduate students, and industrial researchers in these areas of study. It will also cater to researchers and enthusiastic readers in the realm of alternative energy resources as well as in areas of sustainable and green energy technology development.

KEY WORDS

biodiesel, process intensification, sustainable energy, transesterification

CONTENTS

LIST OF FIGURES

LIST OF TABLES

ACKNOWLEDGMENTS

We thank Professor Derek Dunn-Rankin from University of California, Irvine, and Joel Stein, Publisher at Momentum Press for comments on the manuscript and helpful suggestions. This work was supported by a postdoctoral fellowship from the Catalysis and Science Research Center by a research grant from PutraCAT, Faculty of Science, University Putra Malaysia, Malaysia.

CHAPTER 1

INTRODUCTION TO BIODIESEL

The enormous worldwide use of diesel fuel and the rapid depletion of crude oil reserves have prompted keen interest and exhaustive research into suitable alternative fuel. Currently, the world's growing thirst for oil amounts to almost 1000 barrels a second (Armaroli and Balzani 2007). In consequence, the inadequacy of fossil fuel and increases in the demand of energy are the driving forces concerning the future energy security around the world. Thus, the most *important* strategic *issue* of the 21st century states as the struggle for existence is the struggle for sustainable energy. Sustainable energy is energy that has minimal negative impacts on human health and the healthy functioning of vital ecological systems. Currently, attention is focused on human and environmental safety, in relation to the release of hydrocarbons into the environment. Petroleum derivatives contain benzene, toluene, ethylbenzene, and xylene isomers (the major components of fossil fuel), which are hazardous substances subject to regulations in many parts of the world. As a consequence, the demand for green energy is increasingly gaining international attention. When green energy is used, the primary objective is to reduce air pollution and minimize or eradicate completely any impacts to the environment. Among many possible sources, biodiesel is a viable alternative energy to conventional diesel fuel, which is of environmental concern and is under legislative pressure to be replaced by biodegradable substitutes.

Biodiesel refers to the lower alkyl esters of long chain fatty acids (FA), which are synthesized either by transesterification with lower alcohols or by esterification of FA. With minimal subsidy, biodiesel is cost competitive with petroleum diesel, and millions of users have found and enjoyed the benefits of the fuel. The future of biodiesel lies in the world's ability to produce renewable feedstock such as vegetable oils and fats to keep the cost of biodiesel competitive with petroleum, without supplanting land necessary for food production, or destroying natural ecosystems in the process. Creating biodiesel in a sustainable manner will allow this clean, renewable, and cost-effective fuel to help ease the world through increasing shortages of petroleum, while providing economic and environmental benefits well into the 21st century.

Biodiesel is an alternative fuel similar to conventional or fossil diesel. Biodiesel is made from renewable resources like vegetable oil, animal oil or fats, tallow, and waste cooking oil. The concept of using vegetable oil as fuel was introduced by Rudolf Diesel in 1892. Rudolf Diesel developed the first diesel engine that was run with vegetable oil. After eight decades, the awareness about environment rose among the people to search for an alternative fuel that could burn with less pollution. Gerpen and Knothe (2005) reported that a decreased dependence on foreign sources of fuel will enhance national security; thereby, interest in the use of biodiesel as an alternative fuel has accelerated.

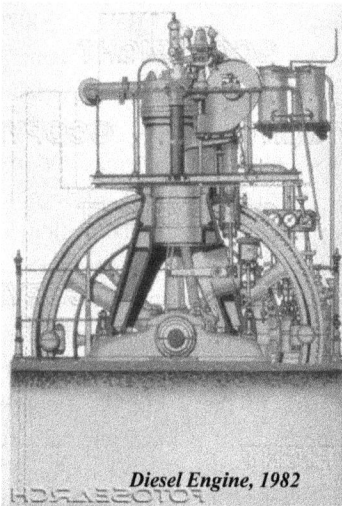

Diesel Engine, 1982   *Rudolph Diesel*

Biodiesel has become more attractive recently because of its environmental benefits and the fact that it is made from renewable resources. The following list includes the key identified advantages of biodiesel, as reported by several researchers (Bajpai and Tyagi 2006; Ganesan, Rajendran, and Thangavelu 2009; Hill et al. 2006; Knothe 2010; Yee et al. 2009):

- Toxicity: Biodiesel is non-toxic, biodegradable, and creates less air pollution than petroleum diesel. It is less toxic than table salt and biodegrades as fast as sugar.
- Biodiesel degrades four times faster than diesel.

- Pure biodiesel degrades 85 to 88 percent in water.
- The blending of biodiesel with diesel fuel increases engine efficiency.
- The higher flash point makes the storage safer.
- Biodiesel is an oxygenated fuel, thus implying that its oxygen content plays a role in making fatty compounds suitable as diesel fuel by "cleaner" burning.
- 90 percent reduction in cancer risks, according to Ames mutagenicity tests.
- Provides a domestic, renewable energy supply.
- Biodiesel does not produce greenhouse effects, because the balance between the amount of CO_2 emissions and the amount of CO_2 absorbed by the plants producing vegetable oil is equal.
- Biodiesel can be used directly in compression ignition engines with no substantial modifications of the engine.
- Biodiesel contains no sulfur and it is also free from sulfur dioxide and aromatic contents benzene, toluene, ethylbenzene, and xylenes (BTEX) emissions that are extremely vigorous toxins in the human body. Acute (short-term) exposure to gasoline and its components benzene, toluene, and xylenes has been associated with skin and sensory irritation, central nervous system (CNS) problems (tiredness, dizziness, headache, loss of coordination), and effects on the respiratory system (eye and nose irritation). Prolonged (chronic) exposure to BTEX compounds can affect the kidney, liver, and blood systems. Long-term exposure to high levels of the benzene compound can lead to leukemia and cancers of the blood-forming organs. It is generally suitable to match the future European regulations that limit the sulfur content to 0.2 percent in weight in 1994 and 0.05 percent in 1996.
- Chemical characteristics revealed lower levels of some toxic and reactive hydrocarbon species when biodiesel fuels were used.
- The emissions of polycyclic aromatic hydrocarbons (PAH) and nitro PAH compounds were substantially lower with biodiesel and are compared to conventional diesel fuel.
- The larger reductions in PAH are not unexpected when considering the biodiesel contains no aromatics and no PAH compounds.
- Biodiesel has a high cetane number (above 100, compared to only 40 for diesel fuel). Cetane number is a measure of a fuel's ignition quality. The high cetane numbers of biodiesel contribute to easy cold starting and low idle noise. The use of biodiesel can extend the life of diesel engines because it is more lubricating and, furthermore, it does not affect the power output.

Despite the many positive characteristics of biofuels, there are also many disadvantages to these energy sources (Bozbas 2008; Demirbas 2007b, 2008a; Melero, Iglesias, and Morales, 2009; Moser 2011):

- A slight increase in NO_2 emissions over petroleum diesel may be experienced, especially with older engines.
- Biodiesel does not store well for periods of time as it will separate, congeal, and in general, break down while in storage.
- Biodiesel is subject to algae growth as water accelerates microbial growth and is naturally more prevalent in biodiesel fuels than in petroleum-based diesel fuels. Care must be taken to remove water from biodiesel during manufacture and from fuel tanks. A special algaecide for diesel can be added to the fuel to inhibit algae growth.
- 100 percent biodiesel and higher percentage biodiesel blends can cause a variety of engine performance problems, including filter plugging, injector choking, piston ring sticking and breaking, elastomer seal swelling and hardening/cracking, and severe engine lubricant degradation.
- Because biodiesel varies based on the manufacturing process and the source materials used, elastomer compatibility with biodiesel remains unclear; therefore, when biodiesel fuels are used, the condition of seals, hoses, gaskets, and wire coatings should be monitored regularly.
- Especially at low ambient temperatures, biodiesel is thicker than conventional diesel fuel, which limits its use in certain geographic areas. This can be solved through the use of winterizing agents also used in petroleum-based diesel fuel or, if practical, you can store the biodiesel in a warm location or heat the fuel tank.
- The biodiesel you use must be free of all foreign material!
- A consortium of diesel fuel injection equipment manufacturers ("FIE Manufacturers") issued a position statement concluding that blends greater than B5 (5 percent biodiesel 95 percent petroleum diesel) can cause reduced product service life and injection equipment failures. According to the FIE Manufacturers' Position Statement, even if the B100 (100 percent biodiesel) is used in a blend and meets one or more specifications, the enhanced care and attention required to maintain the fuels in vehicle tanks may make for a high risk of non-compliance to the standard during use (Nayyar 2010). As a result, the FIE Manufacturers disclaim responsibility for any failures attributable to operating their products with fuels for which the products were not designed.

- There is limited information on the effect of neat (100 percent) biodiesel and biodiesel blends on engine durability during various environmental conditions. More information is needed to assess the viability of using these fuels over the operational life and operating periods typical of heavy-duty engines such as generators.
- Biodiesel is a good solvent and can be corrosive. It will dissolve rubber and some plastics, remove paint, oxidize aluminum, and other metals, and it has been reported to destroy asphalt and concrete if spills were not cleaned quickly. Keep it off items you care about.
- Biodiesels have a lower energy output than traditional fuels and therefore require greater quantities to be consumed in order to produce the same energy level. This has led some noted energy analysts to believe that biofuels are not worth the work.

Food shortage may become an issue with biofuel use (Godfray et al. 2009; Koh and Ghazoul 2008; Singh, Nigam, and Murphy 2011; Walker 2009):

- *Production carbon emissions*: Several studies have been conducted to analyze the carbon footprint of biofuels, and although they may be cleaner to burn, there are strong indications that the process to produce the fuel—including the machinery necessary to cultivate the crops and the plants to produce the fuel—has hefty carbon emissions.
- *High cost*: To refine biofuels to more efficient energy outputs, and to build the necessary manufacturing plants to increase biofuel quantities, a high initial investment is often required.
- *Food prices*: As demand for food crops, such as corn, grows for biofuel production, it could also raise prices necessary for staple food crops.
- *Food shortages*: There is concern that using valuable cropland to grow fuel crops could have an impact on the cost of food and could possibly lead to food shortages.
- *Water use*: Massive quantities of water are required for the proper irrigation of biofuel crops as well as to manufacture the fuel, which could strain local and regional water resources.

REAL CHALLENGE TO BIODIESEL

Biodiesel fuel has gained public appeal for its promise to contribute toward a sustainable energy system and reduce the emission of carbon into the atmosphere. However, there is still much unclarity regarding the effect of the implementation of biodiesel on polluting emissions. One of the most important differences between biodiesel and conventional diesel is the oxygen content. Biodiesel has 10 to 12 percent more oxygen than petroleum-based diesel, which means lower carbon monoxide (CO), particulate matter (PM) emissions, but higher nitrogen oxide (NO_x) emissions and ozone-forming potential as well. Hence, there appears a risk of a steep rise in emissions with the use of biodiesel in particular for NO_x. Therefore, the potential challenge for biodiesel fuel, particularly in the transportation sector, is called "biodiesel NO_x penalty". NO_x or nitric oxides are harmful, toxic, combustion-generated pollution that lead to troposphere ozone, smog, and acid rain. Thus, the following questions may arise:

- Is there a possible win-win scenario, where the implementation of biofuels leads to lower concentrations of, for example, NO_2 or particulate matter?
- Will there be adverse effects, for example, due to an incompatibility of biofuels with modern emission control technology?
- What engine development is expected, for both diesel and petrol engines?
- How does engine technology interact with the use of biofuels, on short and longer term, and what are the expected implications for exhaust emissions?

Several researchers (Agarwal and Das 2001; Choi, Bower, and Reitz 1997; Senatore et al. 2000) have observed increases in NO_x with the use of biodiesel fuel, compared to petroleum diesel. Catalytic converter may be useful to eliminate the NO_2 (Bromberg, Cohn, and Rabinovich 1999; Krahl et al. 2002). With more development and research, it is possible to overcome the disadvantages of biofuels and make them suitable for widespread consumer use. Afterall, biodiesel is not a silver bullet for the energy problems of the world. To solve the issue of dwindling fossil fuel reserves, all viable means of harvesting energy should be pursued to their fullest.

Chemically, biodiesel is defined as the mono-alkyl esters of long chain FA derived from renewable biolipids. The most common way to produce biodiesel is by transesterification, which refers to a catalyzed chemical reaction involving vegetable oil and an alcohol to yield fatty acid alkyl esters (biodiesel) and glycerol, as shown in Figure 1.1. Triglycerides, as the main component of vegetable oil, consist of three long chain FA esterified to a glycerol structure. When triglycerides react with an alcohol (e.g., methanol), the three fatty acid chains are released from the glycerol skeleton and

$$
\begin{array}{l}
R_1COOCH_2 \\
| \\
R_2COOCH \\
| \\
R_3COOCH_2 \\
\textit{Triglyceride}
\end{array}
+ CH_3OH
\xrightleftharpoons{\textit{Catalyst}}
\begin{array}{l}
HOCH_2 \\
| \\
R_2COOCH \\
| \\
R_3COOCH_2 \\
\textit{Diglyceride}
\end{array}
+ R_1COOCH_3
$$

$$
\begin{array}{l}
HOCH_2 \\
| \\
R_2COOCH \\
| \\
R_3COOCH_2 \\
\textit{Diglyceride}
\end{array}
+ CH_3OH
\xrightleftharpoons{\textit{Catalyst}}
\begin{array}{l}
HOCH_2 \\
| \\
HOCH \\
| \\
R_3COOCH_2 \\
\textit{Monoglyceride}
\end{array}
+ R_2COOCH_2
$$

$$
\begin{array}{l}
HOCH_2 \\
| \\
HOCH \\
| \\
R_3COOCH_2 \\
\textit{Monoglyceride}
\end{array}
+ CH_3OH
\xrightleftharpoons{\textit{Catalyst}}
\begin{array}{l}
HOCH \\
| \\
HOCH \\
| \\
HOCH_2
\end{array}
+ R_3COOCH_3
$$

Overall reaction:

$$
\begin{array}{l}
R_1COOCH_2 \\
| \\
R_2COOCH \\
| \\
R_3COOCH_2 \\
\textit{Triglyceride}
\end{array}
+ 3CH_3OH
\xrightleftharpoons{\textit{Catalyst}}
\begin{array}{l}
HOCH_2 \\
| \\
HOCH \\
| \\
HOCH_2 \\
\textit{Glycerol}
\end{array}
\begin{array}{l}
R_1COOCH_3 \\
R_2COOCH_3 \\
R_3COOCH_3 \\
\textit{Methyl ester (Biodiesel)}
\end{array}
$$

$$
RCOOR' + R''OH \xrightleftharpoons{\textit{Catalyst}} R'OH + RCCOR''
$$

$$
\text{Ester} + \text{Alcohol} \xrightleftharpoons{\textit{Catalyst}} \text{Different alcohol} + \text{different ester}
$$

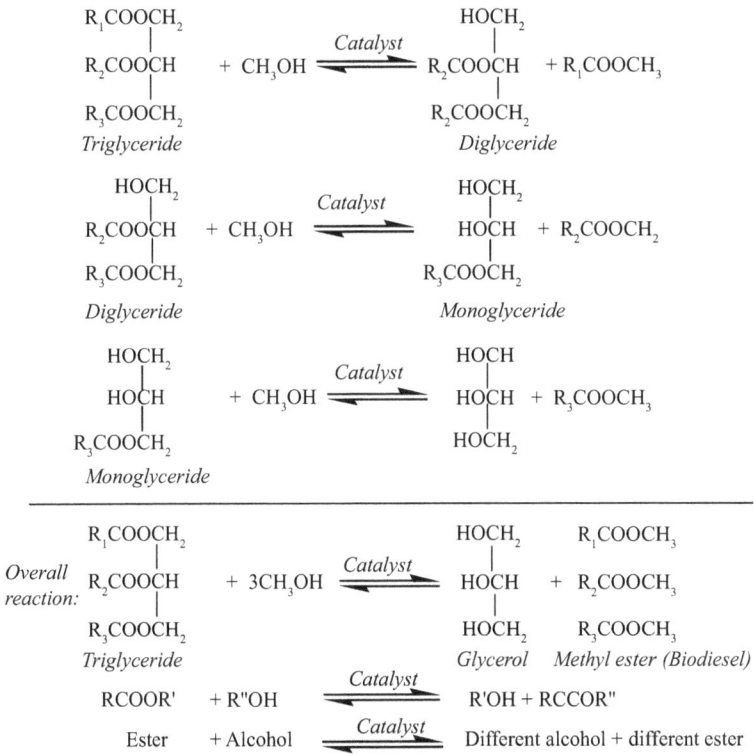

Figure 1.1. Overall transesterification reaction.

combine with the methanol to yield fatty acid methyl esters (FAME). Generally, methanol is preferred for transesterification because it is less expensive than ethanol. The transesterification reaction can be carried out using homogeneous, heterogeneous, or enzymatic catalysts or a non-catalytic process.

Although biodiesel fuel produced from the transesterification of triglycerides contains numerous individual FAME species, a particular fuel is generally dominated by only a few species. A list of FA most commonly seen in biodiesel is provided in Table 1.1.

The dominating composition of FAME derived from vegetable oils and animal fats are palmitic acid (16:0), stearic acid (18:0), oleic acid (18:1), linoleic acid (18:2), and linolenic acid (18:3). Some algal-derived lipids are dominated by these same fatty acid groups, while other algae are more diverse in their composition, containing significant amounts of several other FA groups. Biodiesel (FAME) produced from the transesterification of triglycerides, regardless of their source, is composed nearly exclusively of even-numbered FA chains. In contrast, renewable diesel

Table 1.1. Typical FA groups in biodiesel (Bouaid, Martinez, and Aracil 2007; Hoekman et al. 2012)

Common name	Formal name	Abbreviation	Molecular formula	Molecular structure
Lauric acid	Dodecanoic acid	12:0	$C_{12}H_{24}O_2$	
Myristic acid	Tetradecanoic acid	14:0	$C_{14}H_{28}O_2$	
Myristoleic acid	*cis*-9-Tetradecenoic acid	14:1	$C_{14}H_{26}O_2$	
Palmitic acid	Hexadecanoic acid	16:0	$C_{16}H_{32}O_2$	
Palmitoleic acid	*cis*-9-Hexadecanoic acid	16:1	$C_{16}H_{30}O_2$	
Stearic acid	Octadecanoic acid	18:0	$C_{18}H_{36}O_2$	

Common name	Systematic name	Designation	Formula
Oleic acid	*cis*-9-Octadecenoic acid	18:1	$C_{18}H_{34}O_2$
Linoleic acid	*cis*-9,12-Octadecadienoic acid	18:2	$C_{18}H_{32}O_2$
Linolenic acid	*cis*-9,12,15-Octadecatrienoic acid	18:3	$C_{18}H_{30}O_2$
Arachidic acid	Eicosanoic acid	20:0	$C_{20}H_{40}O_2$
Gondoic acid	*cis*-11-Eicosenoic acid	20:1	$C_{20}H_{38}O_2$
Behenic acid	Docosanoic acid	22:0	$C_{22}H_{44}O_2$
Erucic acid	*cis*-13-Docosenoic acid	22:1	$C_{22}H_{42}O_2$

produced from the same feedstocks contains substantial amounts of odd-numbered FA chains, since one carbon is removed during the hydro-processing step used to manufacture renewable diesel.

1.1 SOURCES OF BIODIESEL

A variety of biolipids can be used to produce biodiesel (Chisti 2007; Demirbas 2008a, 2009b; Karmakar, Karmakar, and Mukherjee 2010). These are (a) virgin vegetable oil feedstock—Camelina, coconut, corn, jatropha, palm, rapeseed, soybean, sunflower, tallow, yellow grease; (b) waste vegetable oil; (c) animal fats including tallow, lard, and yellow grease; and (d) non-edible oils such as jatropha, neem oil, castor oil, tall oil, and so on. There are various other biodiesel sources: almond, andiroba (*Carapa guianensis*), babassu (*Orbignia sp.*), barley, camelina (*Camelina sativa*), coconut, copra, cumaru (*Dipteryx odorata*), *Cynara cardunculus*, fish oil, groundnut, *Jatropha curcas*, karanja (*Pongamia glabra*), laurel, *Lesquerella fendleri*, *Madhuca indica*, microalgae (*Chlorella vulgaris*), oat, piqui (*Caryocar sp.*), poppy seed, rice, rubber seed, sesame, sorghum, tobacco seed, and wheat.

In India and Southeast Asia, the jatropha tree is used as a significant fuel source (Ong et al. 2011). In Malaysia and Indonesia, palm oil is used as a significant biodiesel source (Demirbas 2008a). In Europe, rapeseed is the most common base oil used in biodiesel production. The widespread use of soybeans in the USA for food products has led to the emergence of soybean biodiesel as the primary source for biodiesel in that country. Algae can grow practically anywhere where there is enough sunshine. Vegetable oils are a renewable and potentially inexhaustible source of energy, with energy content close to that of diesel fuel. The common raw materials used for biodiesel production and their oil yield (Oh et al. 2012) are shown in Table 1.2.

1.2 GENERATION OF BIOFUEL

Biofuels are classified into different generations based on the type of feedstocks and technologies used to produce them. First generation biofuels are produced directly from food crops by abstracting the oils for use in biodiesel or producing bioethanol through fermentation. The generation of biofuel was shown in Figure 1.2. There are two main types of first generation biofuels: (a) bioethanol made from sugar cane and corn and (b) biodiesel made from different vegetable oils. Recently, a number of objections have been raised against the use of ethanol produced from

Table 1.2. Common raw materials used for biodiesel production and their oil yield

Oilseeds	Oil yield (ton/ha/year)
Palm oil (Malaysia)	3.93
Rapeseed (EU)	1.33
Soybean (USA)	0.46
Sunflower (Argentina)	0.66
Jatropha (India)	1.44
Coconut (Philippines)	0.66

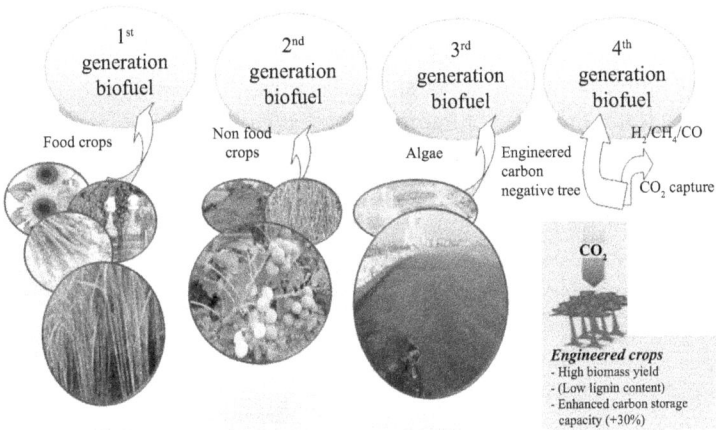

Figure 1.2. Generation of biofuels.

agricultural products such as maize, sugarcane, wheat, or sugar beets as a replacement for gasoline, despite some of their advantages such as being cleaner and to some extent renewable. The "first-generation" biofuels appear unsustainable because of the potential stress that their production places on food commodities (Goldemberg and Guardabassi 2009).

Second generation biofuels are produced from non-food crops such as wood, organic waste, food crop waste, and specific biomass crops, therefore eliminating the main problem with first generation biofuels. Lignocellulosic biomass is an interesting and the necessary enlargement of the biomass used for the production of renewable biofuels. It is expected (Zabaniotou, Ioannidou, and Skoulou 2008) that second generation biofuels are more energy efficient than the ones of the first generation, as a substrate that is able to completely transform into energy.

Third generation biofuel is basically advanced algae-based biodiesel. The algae are cultured to act as a low-cost, high-energy, and entirely renewable feedstock. Microalgae are being promoted as an ideal third generation biofuel feedstock because of their rapid growth rate, CO_2 fixation ability, and high production capacity of lipids; they also do not compete with food or feed crops and can be produced on non-arable land (Dragone et al. 2010). Microalgae have broad bioenergy potential as they can be used to produce liquid transportation and heating fuels, such as biodiesel and bioethanol.

Fourth generation biofuels are produced from genetically modified biomass materials, which have absorbed CO_2 while growing, and are converted into fuel using the same processes as second generation biofuels. The ecological footprint and economic performance of the current suite of biofuel production methods make them insufficient to displace fossil fuels and reduce their impact on the inventory of green house gas (GHG) in the global atmosphere. Lu Jing et al. reported that the algae metabolic engineering forms the basis for fourth generation biofuel production that could reduce GHG (Lü, Sheahan, and Fu 2011). The use of third generation or fourth generation technology, while advanced, should not imply that it is a superior technology in terms of commercial viability. Some of the best early-stage candidates for commercial-scale advanced biofuels are, in fact, second generation biofuel. Feedstock cost, and the capital expense and operating expense of the technology, are major factors in commercial appeal, above and beyond the generation of technology used.

1.3 BIODIESEL STANDARDS SPECIFICATIONS

The nature of the starting material, the production process, subsequent handling, and various factors can influence biodiesel fuel standard. Fuel standard issues are commonly reflected in the contaminants or other minor components of biodiesel. One of the principal means of ensuring satisfactory use in biodiesel fuel quality is the establishment of a rigorous set of fuel specifications. The first ASTM standard (ASTM D6751) was adopted in 2002 (ASTM 2002). Two automotive standards for biodiesel/diesel fuel blends have been published by ASTM:

- The ASTM Standard Specification for Diesel Oil, ASTM D975, was modified in 2008 to allow up to 5 percent biodiesel to be blended into the fuel.
- ASTM D7467 is a specification for biodiesel blends from B6 to B20.

In Europe, EN 14214 establishes specifications for FAME for diesel engines. The US and EU standards have international significance; they are usually the starting point for biodiesel specifications developed in other countries. Numerous other countries have defined their own standards, which in many cases are derived from either ASTM D6751 or EN 14214. Biodiesel specifications and test methods according to US and EU biodiesel specifications (ASTM 2002; Canakci and Sanli 2008; Knothe 2008) are given in Table 1.3.

The US specification, ASTM D6751, defines biodiesel as mono-alkyl esters of long chain FA derived from vegetable oils and animal fats (Moser 2011). The type of alcohol used is not specified. Thus, mono-alkyl esters could be produced with any alcohol (methanol, ethanol, etc.) so long as it meets the detailed requirements outlined in the fuel specification. By requiring that the fuel be mono-alkyl esters of long chain FA, other components, with the exception of additives, would inherently be excluded.

The European biodiesel specification, EN 14214, is more restrictive and applies only to mono-alkyl esters made with *methanol*, FAME. The minimum ester content is specified at 96.5 percent. The addition of components that are not FAME other than additives is not allowed. Guidelines for B100 used to make biodiesel and diesel fuel blends have also been adopted by automobile and engine manufacturers from North and South America, Europe, and Asia (ACEA 2009; Canakci 2008; Knothe 2008). These guidelines bear some resemblance to EN 14214 but there are some notable differences including the following:

- Limiting blends to B5 maximum
- Increasing oxidation stability of B100 to 10 hrs
- Introduction of an oxidation stability requirement for blends that limit the increase in total acid number (TAN) to less than 0.12 mg KOH/g
- Reducing the sulfated ash limit to 0.005 percent from 0.02 percent and introducing an ash limit of 0.001 percent
- Introduction of a ferrous corrosion limit
- Addition of free water and sediment limit
- Loosening of limits for kinematic viscosity, iodine number, and flash point
- Labeling pumps dispensing any blend—including B5 or less

The fatty ester composition, along with the presence of contaminants and minor components, dictates the fuel properties of biodiesel fuel. Biodiesel produced from different feedstocks could have different fuel properties

Table 1.3. US and EU biodiesel specifications

Property	ASTM D975-08a	ASTM D6751-12 2-B	1-B	Test	EN 590:2004	EN 14214:2012
Flash point, min	No 1D 38°C No 2D 52°C	93°C		D93	55°C EN 22719	101°C EN ISO 2719
Water and sediment, max	0.05% vol D2709	0.050% vol		D2709		
Water, max					200 mg/kg EN ISO 12937	500 mg/kg EN ISO 12937
Total contamination, max					24 mg/kg EN ISO 12662	24 mg/kg EN 12662
Distillation temperature (% vol recovered)	90%:1D 288°C max 2D 282–338°C D86	90%: 360°C max		D1160	65%: 250°C min 85%: 350°C max EN ISO 3405	
Kinematic viscosity	1D 1.3–2.4 mm²/s 2D 1.9–4.1 mm²/s D445	1.9–6.0 mm²/s		D445	2.0–4.5 mm²/s EN ISO 3104	3.5–5.0 mm²/s EN ISO 3104

Density					820–845 kg/m³	EN ISO 3675, EN ISO 12185	860–900 kg/m³	EN ISO 3675, EN ISO 12185
Ester content	5% vol. max	EN 14078			5% vol. max FAME	EN 14078	96.5% min	EN 14103
Ash, max	0.01% wt	D482			0.01% wt	EN ISO 6245		
Sulfated ash, max			0.020% mass	D874			0.02% mass	ISO 3987
Sulfur, max (by mass)	1D and 2D: S15 15 mg/kg, S500 0.05%, S5000 0.50%	D5453, D2622, D129[1]	Two grades: S15 15 ppm, S500 0.05%	D5453	Two grades: 50 mg/kg, 10 mg/kg	EN ISO 14596, EN ISO 8754, EN ISO 24269	10.0 mg/kg	EN ISO 20846

(Continued)

Table 1.3. (*Continued*)

Property	ASTM D975-08a		ASTM D6751-12 2-B	1-B	Test	EN 590:2004		EN 14214:2012	
Copper strip corrosion, max	No 3	D130	No 3		D130	Class 1	EN ISO 2160	Class 1	EN ISO 2160
Cetane number, min	40	D613	47		D613	51.0	EN ISO 5165	51.0	EN ISO 5165
Cetane index, min						46.0	EN ISO 4264		
One of²:									
–cetane index	40 min	D976-80							
–aromaticity	35% vol max	D13319							
PAH, max						11% wt	IP 391		
Operability, one of:cloud point and LTFT/CFPP	Report	D2500 D4539 D6371					EN 12916		

Property								
Cloud point			Report	D2500	Location and season dependant	EN 23015	Location and season dependant	EN 23015
CFPP					Location and season dependant	EN 116	Location and season dependant	EN 116
Carbon residue on 10% distillation residue, max	1D: 0.15% wt 2D: 0.35% wt	D524	0.050% wt[3]	D4530	0.30% wt	EN ISO 10370		
Acid number, max			0.50 mg KOH/g	D664			0.50 mg KOH/g	EN 14104
Oxidation stability			3 hrs min	EN 14112	25 g/m³ max	EN ISO 12205	8 hrs min	EN 14112
Iodine value, max							120[4] g Iod/100g	EN 14111 EN 16300

(Continued)

Table 1.3. (*Continued*)

Property	ASTM D975-08a	ASTM D6751-12 2-B	1-B	Test	EN 590:2004	EN 14214:2012
Linolenic acid methyl ester, max						12.0% wt EN 14103
Polyunsatured methyl esters, max						1.00% wt EN 15779
Alcohol control		0.2% wt methanol max, or130°C flash point min		EN14110 D93		0.20% wt methanol max EN 14110
Monoglycerides, diglycerides,and triglycerides, max		MG0.40% wt		D6584		MG 0.70% wt DG 0.20% wt TG 0.20% wt EN 14105
Group I metals (Na + K), max		5 mg/kg		EN 14538		5.0 mg/kg EN 14108 EN 14109 EN 14538

Property						
Group II metals (Ca + Mg), max	5 mg/kg	EN 14538			5.0 mg/kg	EN 14538
Free glycerin, max	0.020% wt	D6584			0.02% wt	EN 14105 / EN 14106
Total glycerin, max	0.240% wt	D6584			0.25% wt	EN 14105
Phosphorous, max	0.001% wt	D4951			4.0 mg/kg	EN 14107 / prEN 16294
Lubricity, max	520 μm	D6079	460 μm	ISO 12156-1		
Conductivity, min	25 pS/m	D2624 / D4308				
Cold soak filtration time (CSFT), max	360 s[5] 200 s	D7501				

[1] D129 is only applicable to S5000 grades.
[2] Limits only apply to S15 and S500 grades.
[3] Tested on 100% sample but reported using 10% residual calculation.
[4] Spain's Royal Decree 1700/2003 sets the maximum iodine value at 140 to facilitate the use of soybean oil as a feedstock.
[5] 200 s if fuel temperature ≤ −12°C.

due to the unique chemical composition of each feedstock. Important properties of biodiesel that are directly influenced by fatty ester composition and the presence of contaminants and minor components include low-temperature operability, oxidative and storage stability, kinematic viscosity, exhaust emissions, cetane number, and energy content.

In the context of biodiesel, minor components (Figure 1.3) are defined as naturally occurring species found in vegetable oils and animal fats and may include steryl glucosides, phospholipids, chlorophyll, fat soluble vitamins, and hydrocarbons (such as alkanes, squalene, carotenes, and PAH) (Giwa, Abdullah, and Adam 2010; Karmakar, Karmakar, and Mukherjee 2010; Moser 2011). Contaminants are defined as incomplete or unwanted reaction products, such as free fatty acids (FFA), soaps, TAG, DAG, MAG, alcohol, catalyst, glycerol, metals, and water (Figure 1.4). It is important to point out that the properties of biodiesel are strongly influenced by the nature of the fatty acid chains in the triglyceride (Sivasamy et al. 2009) such as their length, degree of unsaturation, and presence of other chemical functional groups, which also strongly depends on the source of the feedstock. In general, the cetane number, cloud point, heat of combustion, melting point, and viscosity of the biodiesel increase with increasing chain length of the component fatty acid and decrease with increasing unsaturation (Knothe and Steidley 2005; Sivasamy et al. 2009). In addition, esters with unsaturated fatty acid chains are less stable as a result of relatively fast oxidation.

1.3.1 LOW-TEMPERATURE OPERABILITY

Low-temperature operability of biodiesel is normally determined by three common parameters: cloud point, pour point, and cold filter plugging point (Edith, Janius, and Yunus 2012; Moser 2011). The cloud point of a diesel fuel is the temperature below which wax forms giving the fuel a cloudy appearance. This parameter is an important property of the fuel since the presence of solidified waxes can clog filters and negatively impact engine performance. The cloud point is determined by visually inspecting for a haze in the normally clear fuel. At temperatures below the cloud point, larger crystals fuse together and form agglomerations that eventually become extensive enough to prevent pouring of the fluid. The pour point is the lowest temperature at which a petroleum product will begin to flow. The cold filter plugging point is the temperature at which a fuel filter plugs due to fuel components that have crystallized or gelled. The cold filter plugging point is generally considered to be a more reliable indicator of low-temperature operability than cloud point or pour

Steryl Glucoside (β-sitosterol-β-D-glucopyranoside)

α-Tocopherol R=R′=R′′=H
β-Tocopherol R=R′′=Me; R′=H
γ-Tocopherol R=H; R′=R′′=Me
δ-Tocopherol R=R′=H; R′′=Me

Phospholipid (Phosphatidic Acid)

Carotene (β-carotene)

Squalene

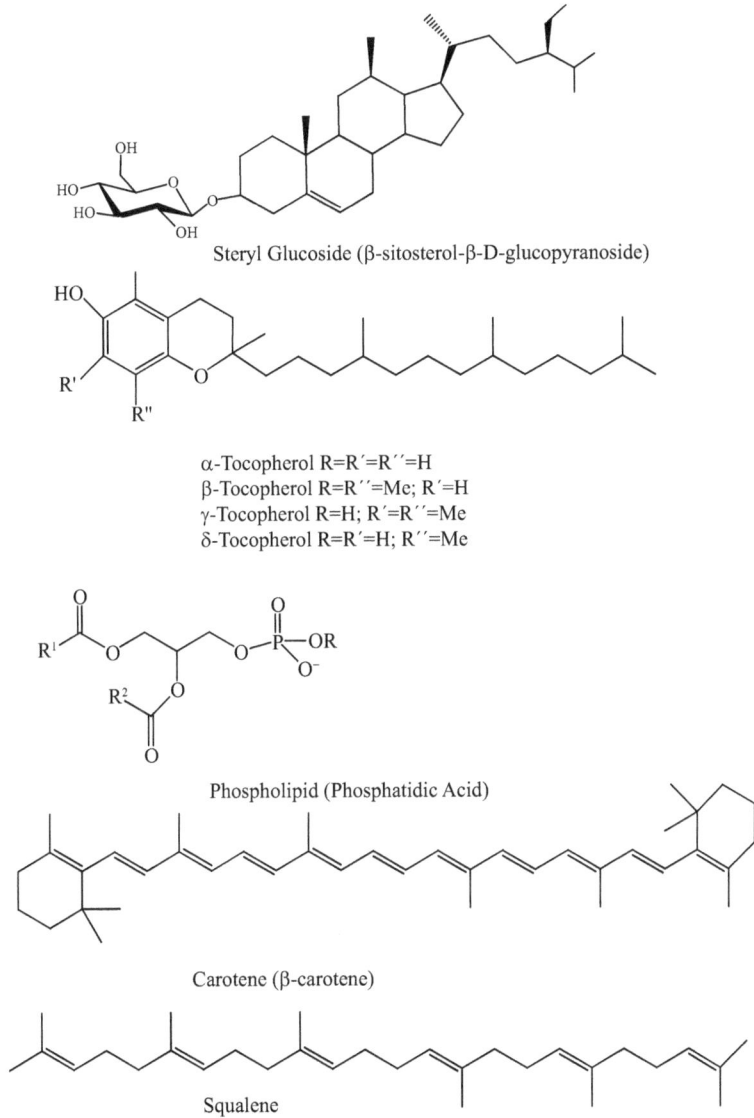

Figure 1.3. Representative examples steryl glucocides, tocopherols, phospholipids, and hydrocarbons that may be found in biodiesel.

point, because the fuel will contain solids of sufficient size to render the engine inoperable due to fuel filter plugging once the cold filter plugging point is reached (Dunn and Bagby 1995; Dunn, Shockley, and Bagby 1996; Knothe 2005). It should be noted that it is inappropriate to measure cloud point, pour point, and cold filter of chemically pure compounds

Figure 1.4. Soap formation in transesterification reaction.

(pure methyl oleate, for instance). Instead, the determination of melting point as a means to measure low-temperature operability is appropriate for chemically pure compounds. The low-temperature behavior of chemical compounds is dictated by molecular structure. Structural features such as chain length, degree of unsaturation, orientation of double bonds, and type of ester head group strongly influence the individual chemical constituents of biodiesel.

1.3.2 OXIDATIVE STABILITY

Oxidation stability is a chemical reaction that occurs with a combination of the lubricating oil and oxygen. The rate of oxidation is accelerated by high temperatures, water, acids, and catalysts such as copper. The rate of oxidation increases with time. The service life of a lubricant is also reduced with increases in temperature. Oxidation will lead to an increase in the oil's viscosity and deposits of varnish and sludge. The oxidative stability of biodiesel is determined through the measurement of the oil stability index by the Rancimat method (Cavalcanti et al. 2011). The Rancimat method indirectly measures oxidation by monitoring the gradual change in conductivity of a solution of water caused by volatile oxidative degradation products that have been transported via a stream of air (10 L/h) from the vessel (at 110°C) containing the biodiesel sample. Oxidative stability and low-temperature operability are normally inversely related: structural factors that improve oxidative stability adversely influence low-temperature operability and vice versa (Hada, Solvason, and Eden 2014).

1.3.3 KINEMATIC VISCOSITY

Kinematic viscosity of oil is its resistance to flow at a specific temperature. The viscosity of a fuel decreases with increasing temperature. Kinematic viscosity is the primary reason why biodiesel is used as an alternative fuel instead of neat vegetable oils or animal fats. High viscosity can cause larger droplet sizes, poorer vaporization, narrower injection spray angle, and greater in-cylinder penetration of the fuel spray (Hoekman et al. 2012; Zeng et al. 2010; Basinger et al. 2010). This can lead to overall poorer combustion, higher emissions, and increased oil dilution. The high kinematic viscosities of vegetable oils and animal fats ultimately lead to operational problems such as engine deposits when used directly as fuels (Knothe and Steidley 2005, 2007, 2010). The kinematic viscosity of biodiesel is approximately an order of magnitude less than typical vegetable oils or animal fats and is slightly higher than petrodiesel. The methyl and ethyl esters of canola oil have kinematic viscosities of 3.9 and 4.4 mm^2/s, respectively (Ejigu et al. 2010).

1.3.4 EXHAUST EMISSIONS

Biodiesel is the only alternative fuel to successfully complete the Environmental Protection Agency's (EPA's) rigorous emissions and health effects study under the Clean Air Act (Yamane, Ueta, and Shimamoto 2001). Biodiesel provides significantly reduced emissions of carbon monoxide, particulate matter, unburned hydrocarbons, and sulfates compared to petroleum diesel fuel. Additionally, biodiesel reduces emissions of carcinogenic compounds by as much as 85 percent compared with petrodiesel. When blended with petroleum diesel fuel, these emission reductions are generally directly proportional to the amount of biodiesel in the blend. The exhaust emissions of particulate matter from biodiesel are about 47 percent lower than overall particulate matter emissions from diesel (Kwanchareon, Luengnarucmitchai, and Jai-In 2007). The EPA (Mavournin et al. 1990) reported that biodiesel decreases the tailpipe emissions of PM, CO, and hydrocarbons (HC) commensurately with its blend level, as shown in Figure 1.5. Biodiesel helps reduce the risk of global warming by reducing net carbon emissions to the atmosphere. When biodiesel is burned, it releases carbon dioxide to the atmosphere, but crops that are used to produce biodiesel take up carbon dioxide from the atmosphere in their growth cycle.

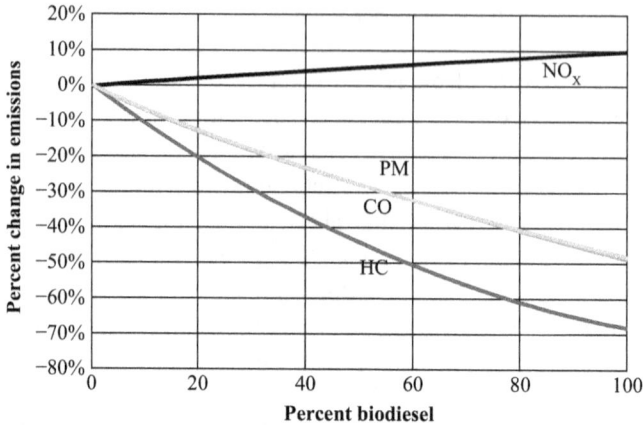

Figure 1.5. Average emission impact of biodiesel for heavy-duty highway engines.

1.3.5 CETANE NUMBER

Cetane number is determined in accordance with ASTM D613 and is one of the primary indicators of diesel fuel quality. Cetane number is a measure of a fuel's autocombustion quality. It refers the time delay between the start of injection process and the point where the fuel ignites. This value is determined by the percent volume of cetane that provides the identical ignition delay of the measured fuel sample. Generally, shorter ignition delay times result in higher Cetane number and vice versa (Xing-cai et al. 2004). Hexadecane, also known as cetane (trivial name), which gives the cetane scale its name, is the high-quality reference standard with a short ignition delay time and an arbitrarily assigned CN of 100. It was reported that the cetane index (ASTM D976 or D4737) is not applicable to biodiesel (Moser 2011). The cetane index is used in the case of middle distillate fuels (i.e., ultra-low sulfur diesel (ULSD)) as an estimation of CN and is calculated from density, API gravity, and boiling range (Murphy, Taylor, and McCormick 2004). However, biodiesel has dramatically different distillation qualities (e.g., much higher boiling range) than diesel fuels, thus rendering the equation used to calculate the cetane index inapplicable to biodiesel.

1.3.6 HEAT OF COMBUSTION

Heat of combustion is the thermal energy that is liberated upon combustion, so it is commonly referred to as energy content. Factors that influence the energy content of biodiesel include the oxygen content and carbon to hydrogen ratio. Due to its high oxygen content, biodiesel has lower mass energy values than petroleum diesel. Therefore, increasing the B-level of biodiesel blends results in decreasing energy content. As the fatty acid carbon chain increases, the mass fraction of oxygen decreases, so the heating value increases (Van Gerpen and He 2010). Generally, as the oxygen content of FAME is increased, a corresponding reduction in energy content is observed.

1.3.7 LUBRICITY

Lubricity refers to the reduction of friction between solid surfaces in relative motion (Van Gerpen and He 2010). Two general mechanisms contribute to overall lubricity: (1) hydrodynamic lubrication and (2) boundary lubrication. In hydrodynamic lubrication, a liquid layer (such as diesel fuel within a fuel injector) prevents contact between opposing surfaces. Boundary lubricants are compounds that adhere to the metallic surfaces, forming a thin, protective, anti-wear layer. Boundary lubrication becomes important when the hydrodynamic lubricant has been squeezed out or otherwise removed from between the opposing surfaces. Shorter wear scar values indicate that the sample has superior lubricity versus another sample that resulted in a longer wear scar. The petrodiesel standards, ASTM D975 and EN 590, contain maximum allowable wear scar limits of 520 and 460 μm, respectively (Sharma et al. 2014). Biodiesel possesses inherently good lubricity, especially when compared to petrodiesel (Margaroni 1998; Sharma et al. 2014).

1.3.8 CONTAMINANTS

Contaminants in biodiesel may include methanol, water, catalyst, glycerol, FFA, soaps, and metals. The effect of contaminants on biodiesel and engines reported by several researchers (Canakci and Van Gerpen 1999; Sarin et al. 2009; Van Gerpen et al. 1997) is shown in Table 1.4. Methanol contamination in biodiesel is indirectly measured through flash point determination following ASTM D93. If biodiesel is contaminated with methanol, it will fail to meet the minimum flash point specified in relevant fuel standards. Methanol contamination normally results from the insufficient purification of biodiesel following the transesterification reaction. Water is a major source of fuel contamination. While fuel leaving a production facility may be virtually free of water, once it enters the existing

Table 1.4. Negative effects of contaminants on biodiesel and engines

Contaminants	Negative effect
Methanol	Deterioration of natural rubber seals and gaskets, lower flash points (problems in storage, transport, utilization, etc.), lower viscosity and density values, corrosion of pieces of aluminum (Al) and zinc (Zn)
Water	Reduces heat of combustion, corrosion of system components (such as fuel tubes and injector pumps) failure of fuel pump, hydrolysis (FFAs formation), formation of ice crystals resulting to gelling of residual fuel, bacteriological growth causing blockage of filters, and pitting in the pistons
Catalyst/soap	Damage injectors, pose corrosion problems in engines, plugging of filters, and weakening of engines
FFAs	Less oxidation stability, corrosion of vital engine components
Glycerides	Crystallization, turbidity, higher viscosities, and deposits formation at pistons, valves, and injection nozzles
Glycerol	Decantation, storage problem, fuel tank bottom deposits, injector fouling, settling problems, higher aldehydes and acrolein emissions, and severity of engine durability problems

distribution and storage network it will come into contact with water as a result of environmental humidity (Knothe et al. 2005).

Minor components in biodiesel may include copherols, phospholipids, steryl glucosides, chlorophyll, fat soluble vitamins, and hydrocarbons (Moser 2011). The quantities of these components depend on the feedstock from which the biodiesel is prepared, how the biodiesel is purified, and the degree of pre-processing (refining, bleaching, deodorization, degumming, etc.) that is performed on the feedstock prior to transesterification.

1.4 BIODIESEL AS TRANSPORTATION FUEL

The world's current transportation systems are highly dependent on petroleum, a resource that is concentrated in relatively few countries. This has left the global economy at risk of disruption, particularly with oil supplies as tight as they are now. Biofuels promise to bring a much broader group of countries into the liquid fuel business, diversifying supplies and reducing the risk of disruption. Of the world's 47 poorest countries, 38 are net oil importers, and 25 of these import all of their oil. In many smaller and poorer nations, 90 percent or more of the total energy used comes from imported fossil fuels (Hunt 2007). In some cases, a large share of the foreign exchange earnings goes to pay for oil, and much of the government revenue is used to subsidize kerosene and diesel fuel. Yet, many of these same countries have substantial agricultural bases and are well suited to growing sugar cane, palm oil, and other highly productive energy crops. Some of these countries even have the potential to become net exporters of liquid fuels. International trade in biofuels is currently limited by the fact that many countries maintain tariffs on these fuels, both to protect their domestic industries and to assure that their substantial domestic subsidies are not used to support the industries of other nations. This is likely to change in the years ahead. Many of the rich countries that consume large quantities of transportation fuels (e.g., Europe and Japan) have limited land available for growing biomass feedstock, which leaves them unable to generate more than a fraction of their transportation fuels from domestically produced biofuels (Molitor et al. 2007).

Some countries may decide to eliminate biofuel tariffs on a bilateral basis with individual trading partners. The United States, for example, already allows the preferential import of ethanol from the Caribbean. Sweden wants to encourage large-scale biofuel imports. Ongoing negotiations at the World Trade Organization, aimed at liberalizing trade in agricultural commodities, are expected to address the potential for reducing biofuel trade barriers, offering an opportunity for countries to generate new agricultural revenue streams to offset the loss of trade-distorting subsidies.

It was reported by several researchers (Bomb et al. 2007; Demirbas 2011) that biodiesel is a processed fuel that can be readily used in diesel-engine vehicles, which distinguishes biodiesel from the straight vegetable oils or waste vegetable oils used as fuels in some modified diesel vehicles. In an experimental setup performed by Chattopadhyay and Sen (2013) for fuel properties, the engine performance testing of biodiesel was shown in Figure 1.6.

The advantages of biodiesel as diesel fuel are its portability, ready availability, renewability, higher combustion efficiency, and lower sulfur and aromatic content, higher cetane number, and higher biodegradability. The main advantages of biodiesel given in the literature (Mekhilef, Siga, and Saidur 2011; Pascoli, Femia, and Luzzati 2001; Van de Velde et al.

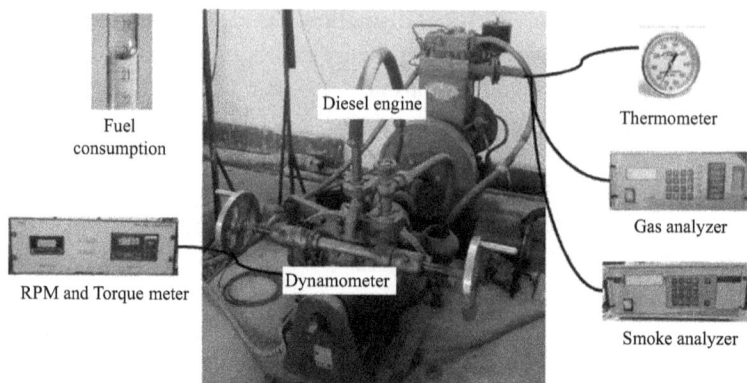

Figure 1.6. An experimental setup for fuel properties, engine performance testing of biodiesel.

2009) include its domestic origin, which would help reduce a country's dependency on imported petroleum, its biodegradability, high flash point, and inherent lubricity in the neat form. The major disadvantages of biodiesel are its higher viscosity, lower energy content, higher cloud point and pour point, higher NOx emissions, lower engine speed and power, injector coking, engine compatibility, high price, and greater engine wear. The technical disadvantages of biodiesel/fossil diesel blends include problems with fuel freezing in cold weather, reduced energy density, and degradation of fuel under storage for prolonged periods.

A large number of studies (Zhang et al. 2003a; Haas et al. 2006; Marchetti, Miguel, and Errazu 2008) have been performed over the last decade to assess the economic feasibility of biodiesel and reported that the costs for biodiesel from oilseed or animal fats have a range of $0.30 to 0.69/l, including the meal and glycerin credits and the assumption of reduced capital investment costs by having the crushing and esterification facility added onto an existing grain or tallow facility. Rough projections of the cost of biodiesel from vegetable oil and waste grease are, respectively, $0.54 to 0.62/l and $0.34 to 0.42/l. With pre-tax diesel priced at $0.18/l in the United States and $0.20 to 0.24/l in some European countries, biodiesel is, thus, currently not economically feasible, and more research and technological development will be needed (Demirbas 2003, 2006). One potential solution to this problem is employment of alternative feedstocks. These feedstocks may include soapstocks, acid oils, tall oils, used cooking oils, and waste restaurant greases, various animal fats, nonfood vegetable oils, and oils obtained from trees and micro-organisms such as algae. However, many of these alternative feedstocks may contain high levels of FFA, water, or insoluble matter, which affect biodiesel production.

Skyrocketing prices of crude oil in the middle of the first decade of the 21st century accompanied by rising prices for food focused political and public attention on the role of biofuels. On the one hand, biofuels were considered as a potential automotive fuel with a bright future, and on the other hand, biofuels were accused of competing with food production for land. For biofuels to make a large and sustainable contribution to the world energy economy, governments will need to enact consistent, long-range, and coordinated policies that are informed by broad stakeholder participation. Several researchers (Gadonneix et al. 2010; Garcez and Vianna 2009; Körbitz 2009; Leite et al. 2013; McMichael 2012; Tilman et al. 2009) suggested a set of *recommendations* for moving *biofuel production* toward being more sustainable. Recommendations include the following:

- *Strengthen the market*: Biofuel policies should focus on market development, creating an enabling environment based on sound fiscal policy and support for private investment, infrastructure development, and the building of transportation fleets that are able to use the new fuels.
- *Speed the transition to next-generation technologies*: Policies are needed to expedite the transition to the next generation of feedstock and technologies that will enable dramatically increased production at lower cost, while reducing negative environmental impacts.
- *Protect the resource base*: Maintaining soil productivity, water quality, and myriad other ecosystem services is essential. National and international environmental sustainability principles and certification systems are important for protecting resources as well as maintaining public trust in the merits of biofuels.
- Encourage broad rural economic benefits.
- Government fiscal and land use policies will help determine how broadly the economic revenues from biofuels are spread and how they will shape rural economies.
- *Facilitate Sustainable International Biofuel Trade:* Continued rapid growth of biofuels will require the development of a true international market in the fuels, unimpeded by the trade restrictions in place today. Freer movement of biofuels around the world should be coupled with social and environmental standards and a credible system to certify compliance.
- *Efficiency and improved public transport*: Biofuels should be developed within the context of a broad transformation of the transport sector aimed at dramatically improving transport efficiency.

- Supportive government policies have been essential to the development of modern biofuels over the past two decades. Countries seeking to develop domestic biofuel industries will be able to draw important lessons both positive and negative from the industry pioneers: Brazil, the United States, and the European Union. The following are among the successful policies that have fostered biofuel production and use:

  - Blending Mandates
  - Tax Incentives
  - Government Purchasing Policies
  - Support for Biofuel-Compatible Infrastructure and Technologies
  - R&D (including crop research, conversion technology development, feedstock handling, etc.)
  - Public Education and Outreach
  - Reduction of Counterproductive Subsidies
  - Investment Risk Reduction for Next-Generation Facilities
  - Gradual Reduction of Supports as the Market Matures

The nature and the magnitude of biofuel impacts depend on the feedstock. A more sustainable biofuels strategy would be a utilization of widely available agricultural biomass feedstocks to the largest extent possible, ultimately drawing upon lignocellulosic biomass instead of only the edible oils and triglyceride fractions. The first generation technologies have drawbacks in that they rely on feedstocks that are not sufficiently available to satisfy the demands presently met by petroleum, and they rely on easily accessible edible biomass fractions, thereby impacting the supply of food for humans and animals. Thus, the development of second and third generation biofuels that utilize lignocellulosic biomass and algae could be a better option to allow large-scale production of sustainable biofuels. It is worth mentioning that the waste carbon dioxide exists in plenty and causes a negative impact on the environment by including contributing to the greenhouse effect. With this in mind, it would be attractive to utilize waste carbon dioxide as a future unlimited feedstock for biodiesel production that can be employed as drop-in replacements for fossil fuels.

CHAPTER 2

REACTOR AND REFINING TECHNOLOGIES FOR BIODIESEL PRODUCTION

Three general types of reactors are used for biodiesel production: batch reactors, semi-continuous-flow reactors, and continuous-flow reactors. The batch process is inexpensive, requiring much less initial capital and infrastructure investment. It is flexible and allows the user to accommodate variations in feedstock type, composition, and quantity. The major inconvenience of batch process lays on the low productivity, larger variation in product quality, and more intensive labor and energy requirements (Behzadi and Farid 2009; Nabeel Adedapo, Mohiuddin, and Jameel 2011). The semi-continuous process is similar to the batch process except that the producer starts by reacting a smaller volume than the vessel will hold and then continues to add ingredients until the vessel is full. This process is labor intensive and not commonly used. Continuous transesterification processes are preferred over batch processes in large capacity commercial production because these processes result in consistent product quality and low capital and operating costs per unit of product. The most common type of continuous-flow reactor is the continuous stirred-tank reactor. Other types of continuous-flow reactors are also used commercially, including ultrasonic reactors and supercritical reactors.

Building a sustainable biodiesel industry, extra efforts are still required in the research and development of biodiesel production by developing a robust technology to produce, refine, and recover the valuable end products. In order to thoroughly understand the development of biodiesel production technology, it is indeed important to assessment the conventional technologies as well as to have an understanding of the recent advances in biodiesel production technology. Considering the current drawbacks in conventional biodiesel production technologies, new technologies have been developed by the collaboration of many researchers and industries based on the process intensification (PI) approach. The definition of PI was given by Reay (2008) that any chemical engineering development that leads to a substantially smaller, cleaner, and more energy-efficient technology is PI. According to Heimann (2003), PI could improve yield or selectivity and facilitate separation, thus resulting in a lower investment cost, smaller inventory (safety aspects), and improved heat management/ energy utilization. It was reported by Keil (2007) that a better design of the reactor can improve the economics of biodiesel facilities by significantly reducing capital and operating costs, improving flexibility to process low-cost feedstock, and improving yields of biodiesel of a quality that satisfies international standards.

Separation of glycerol from the reaction mixture is usually quick due to the difference in polarities and larger density difference between glycerol and esters. The separation is usually accomplished through gravitational settling or centrifugation. The crude biodiesel obtained after the separation of glycerol is subjected to either distillation or rotary evaporation in order to remove the residual alcohol. The process of biodiesel water washing usually provides a final biodiesel product that satisfies the stringent biodiesel standards such as EN 14214 and ASTM D6751. However, separation of hot wash water and the acid from biodiesel in some cases requires the application of two centrifuges. Besides, the water washing process generates wastewater containing impurities such as free glycerol, residual catalyst, and soap, which must be carefully handled before being discharged (Atadashi et al. 2012b), otherwise the disposal of biodiesel wastewater could negatively affect the environment. The problems associated with water washing have led to the use of dry washing process to purify crude biodiesel. Although the process of dry washing with magnesol, or ion exchange resins, provides biodiesel with low impurities and good physicochemical properties, the lack of adsorbents reusability and little knowledge about the chemistry of the adsorbents discourage their use (Atadashi et al. 2012b). Presently, membrane technology is being explored and exploited for the purification of biodiesel. Generally, membrane processes play a critical role in purifying biotechnology products (Dubé, Tremblay, and Liu 2007). Membrane-based separations are well-established technologies in protein separations, water purification, and gas separations. The use of membrane separation for the treatment of non-aqueous fluids is an emerging field in membrane technologies (Atadashi, Aroua, and Aziz 2011a; Atadashi et al. 2011; Shuit et al. 2012). Membranes can be inorganic or organic (polymeric) in nature. The inorganic membranes, particularly ceramic membranes, are more appropriate for use with organic solvents because of their excellent thermal and chemical stability and their ability to withstand higher temperatures and pressures (Wang et al. 2009; Atadashi, Aroua and Aziz 2011; Shuit et al. 2012). Further, the exceptional physical and chemical stability of ceramic membranes permits them to offer reproducible performance over long service life-times, which is well proven in many industrial installations. The conventional biodiesel production technologies as well as detail knowledge of the recently emerged PI technologies for biodiesel production and purification technologies will be described in subsequent sections.

2.1 REACTOR TECHNOLOGIES

2.1.1 BATCH REACTORS

Biodiesel is mainly produced in batch chemical reactors. This is where the oil/fat is placed in a temperature-controlled, stirred tank and reagents such as alcohol and catalyst are added sequentially. The reaction is then stirred for a period of time before draining the contents of the reactor. A typical example of transesterification batch reaction apparatus was shown in Figure 2.1. The major drawbacks of batch reactor are the requirement of large reactor size, the need of high energy consumption, and the high cost for operation. In addition, the quality of the product in each batch reactor is difficult to control (Ma and Hanna 1999; Gerpen 2005). In order to reduce the batch reactor process problems, the researchers have placed more attention to produce biodiesel using continuous process.

2.1.2 CONTINUOUS-FLOW REACTORS

The most common continuous-flow system in biodiesel production is the continuous stirred-tank reactor. A laboratory-scale continuous-flow reactor system was shown in Figure 2.2. Some continuous-flow plants may be able to operate in either batch or continuous mode. In a continuous process, the reactants are continuously added and the product (mixture of different chemicals, including unreacted reactants) continuously withdrawn. Adequate agitation is required to ensure uniform chemical composition and temperature. The continuous-flow process typically requires intricate process controls and online monitoring of product quality. When a reactor is operated continuously at a steady state, ideally the concentration of any chemical involved should be approximately constant anywhere in the reac-

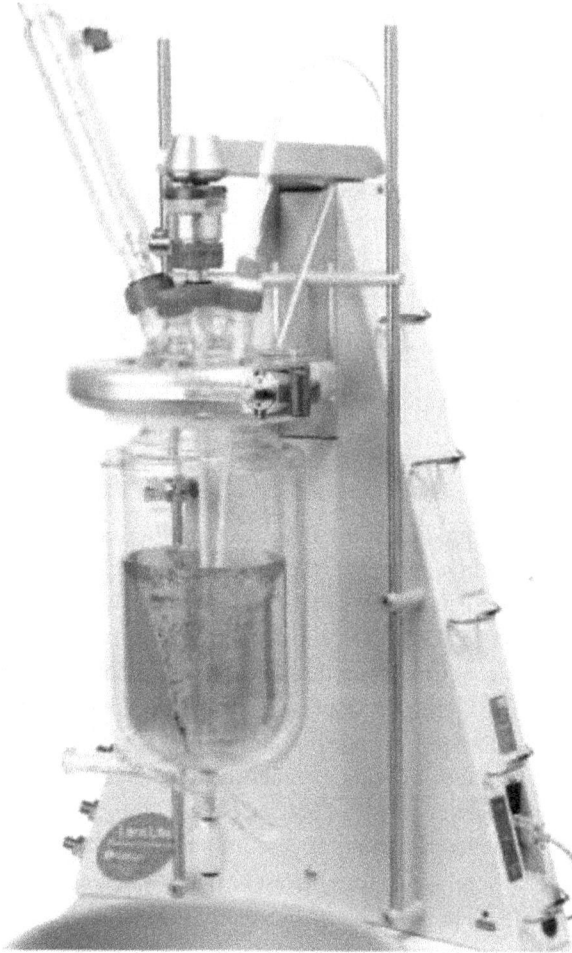

Figure 2.1. Batch transesterification reactor.

tor and at all times. The excess alcohol remaining in the esters and glycerol after reaction must be recovered and purified for reuse. Continuous transesterification processes are preferred over batch processes in large-capacity commercial production due to the consideration of consistent product quality and low costs of capital and operation per unit of product. Continuous transesterification of vegetable oils to mono-alkyl esters was proposed as early as in the 1940s (Russell 1945) and studied until recent years (Harvey, Mackley, and Seliger 2003; Peterson et al. 2002; McNeff et al. 2008; Knothe 2005). Most of the existing processes still utilize 100 percent (by molar quantity) or more excess alcohol. The operating parameters, such as

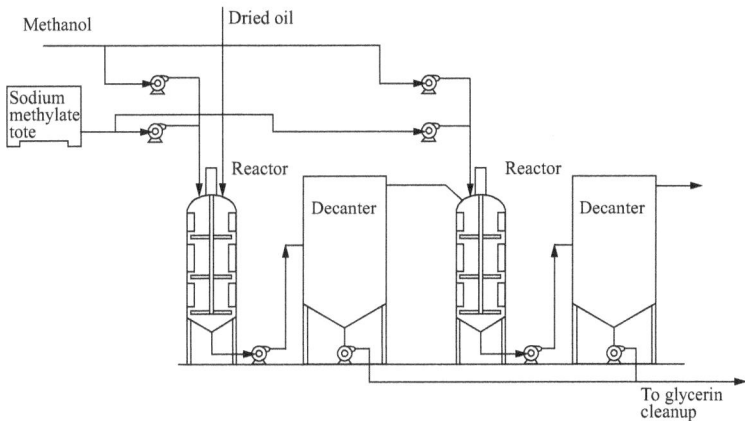

Figure 2.2. Continuous-flow biodiesel reactor two-step process.

reaction time of 60 to 120 min and operating temperatures of 20°C to 70°C, were directly adopted from batch operation processes.

2.1.3 REACTOR COMBINED WITH REACTIVE DISTILLATION COLUMN

Reactive distillation (RD) is a chemical unit operation in which chemical reactions and product separations occur simultaneously in one unit. It is an effective alternative to the classic combination of reactor and separation units. A novel reactor system using RD, as shown in Figure 2.3, was developed and investigated by He, Singh, and Thompson (2006) for biodiesel production from canola oil and MeOH. An RD system consists of numerous chambers with openings from one to the next. Ingredients are added to the first chamber, and as the mixture enters each successive chamber, the reaction progresses so that by the last chamber, the reaction is completed. Both packed and tray columns may be used for RD applications; however, tray columns are preferred for homogeneous reaction systems because of the greater liquid holdup and the relatively longer retention time. It can be noted that the difference between the boiling temperatures of methanol and fatty acid esters (biodiesel) is so large that the separation of these two streams becomes very easy. Because the transesterification reaction occurs in the liquid phase only, the reaction time is then established by the total liquid holdup and the feeding rate of the reactants. According to He, Singh, and Thompson (2005, 2006, 2007), the RD reactor system showed three major advantages over the batch and traditional continuous-flow

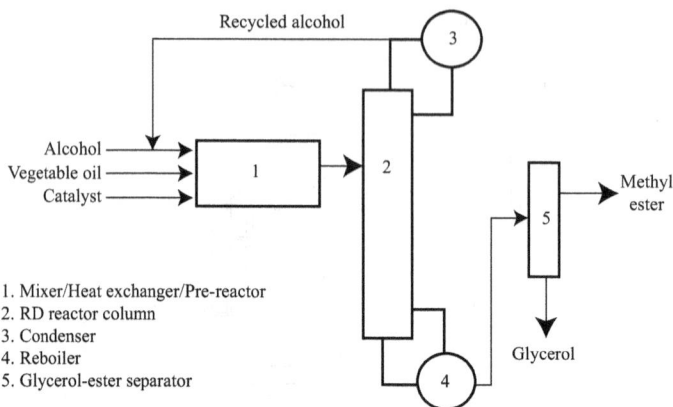

Figure 2.3. Biodiesel production process with distillation column.

processes: (1) shorter reaction time (10 to 15 min) and higher unit productivity (7 to 9 gallons per gallon reactor volume per hour), which is highly desirable in commercial production units; (2) much lower excess alcohol requirement (approximately 3.5:1 molar), which greatly reduces the effort of downstream alcohol recovery and operating costs; and (3) lower capital costs due to its smaller size and the reduced need for alcohol recovery equipment.

2.1.4 ULTRASONIC BIODIESEL REACTORS

In biodiesel production, vigorous mixing is required to create sufficient contact between the vegetable oil/animal fat and alcohol, especially at the beginning of the reaction. "Ultrasound" refers to sound waves that are above the frequency for human hearing, which is approximately 20 kilohertz (kHz), or 20,000 cycles per second. Ultrasound is a useful tool to mix liquids that tend to separate. Ultrasonic waves cause intense mixing at micro-levels and improve mass transfer greatly, so that the reaction can proceed at a much faster rate (Nazir et al. 2009; Cintas et al. 2010). Although not currently in wide use, ultrasound is a promising technology for biodiesel production. Ultrasound processing results in similar yields of biodiesel with a much shortened reaction time compared to the conventional stirred-tank procedure. Ultrasonic reactors can process triglycerides into biodiesel within minutes. In addition, current users of the technology claim that much less catalyst and methanol are required (Kumar et al. 2010a; Veljković, Avramović, and Stamenković 2012).

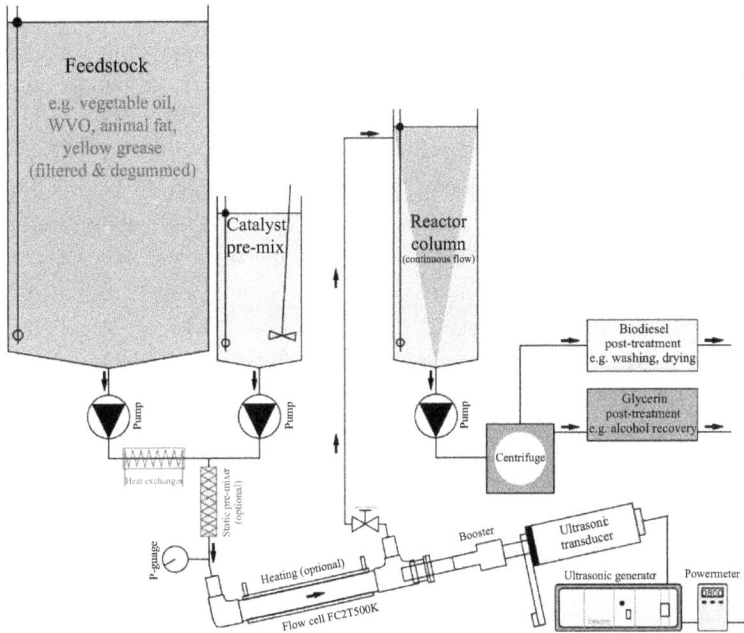

Figure 2.4. Continuous ultrasonic biodiesel processing and separation developed by Hielsher GmbH.

According to Hielscher Ultrasonics GmbH, continuous ultrasonic biodiesel processing and separation can be achieved by first mixing the heated oil with the pre-mixed alcohol and catalyst continuously as shown in Figure 2.4. The mixture is directed through the flow cell, where it will be exposed to ultrasonic cavitation for about 5 to 30 s. After that, the sonicated mixture enters the reactor column that is designed to give approximately 1 hour of retention time for the transesterification reaction to be completed. The biodiesel and glycerine products were separated in the centrifuge. It is then followed by the continuous products' post-treatment steps for alcohol recovery, product washing, and drying. The shortened reaction time and increased biodiesel production could be due to the following reasons, as reported by several researchers (Hanh et al. 2009a, 2009b; Mootabadi et al. 2010; Yu et al. 2010):

- The rapidly vibrating ultrasound waves transfer energy into the fluid and create violent vibrations, which form "cavitation" bubbles as the low pressure part of the sound passes through the liquid. After the wave passes, the bubbles collapse, causing a sudden contraction of the fluid. This collapse produces very intense mixing in the area of the bubbles. Such a high-energy action in the liquid can considerably increase the reactivity of the reactant mixture and shorten the reaction time without involving elevated temperatures. In fact, this reaction can be achieved at or slightly above ambient temperature.

- Ultrasound provides mechanical energy for mixing in which the microturbulence generated due to radial motion of bubbles leading to intimate mixing of the immiscible reactants and thus, initiating the transesterification reaction.
- The ultrasonic irradiation of a liquid produces acoustic cavitations in which H^+ and OH^- are produced during a transient implosive collapse of bubbles that could accelerate the reaction rate.
- Ultrasonic can also grind the catalyst into smaller particles to create new active sites for the subsequent reaction. Thus, the solid catalyst is expected to last longer in the ultrasonic-assisted process.

The ultrasound processing results in similar yields of biodiesel with a much shortened reaction time compared to the conventional stirred-tank procedure. Ultrasound can be a good choice for small producers (up to 2 million gallons per year capacity), who may only need one or two ultrasound probes per reactor vessel. However, using ultrasound in large-scale processing may be challenging because many ultrasound probes would be needed to reach every area of the reactant mixture.

2.1.5 MICROWAVE REACTOR

Recently, microwave reactors have been developed for biodiesel synthesis. The mixture of vegetable oil, methanol, and alcohol contains both polar and ionic components, and microwave irradiation can play an active role in heating reactants to the required temperature quickly and efficiently (Gude et al. 2013; Leadbeater and Stencel 2006). According to Gude et al. (2013), microwave heating offers several advantages over conventional heating such as non-contact heating (reduction of overheating of material surfaces), energy transfer instead of heat transfer (penetrative radiation), reduced thermal gradients, material selective and volumetric heating, fast start-up and stopping, and reverse thermal effect, that is, heat starts from the interior of material body. In terms of biodiesel production, the resultant value could include the following: more effective heating, fast heating of catalysts, reduced equipment size, faster response to process heating control, faster start-up, increased production, and elimination of process steps. The microwave reactor design for biodiesel production was proposed by Gude et al. (2013) and is shown in Figure 2.5. Microwaves at 915 MHz (used industrially) have much higher penetration depths into the material as compared to the higher frequency of 2450 MHz commonly used in laboratory-sized equipment. The higher penetration depths allow for much larger diameter tubes and processing flow rates, and microwave generators can be built for significantly higher power and efficiencies when compared to smaller generators.

A conclusion has been reached by Kumar, Kumar, and Chandrashekar (2011) that the reaction time of microwave-assisted transesterification was shorter compared to conventional methods. Similarly, Yuan, Yang, and Zhu (2008) reported that the reaction time of biodiesel synthesis using microwave irradiation was decreased by 180 min in comparison with conventional heating under the same reaction conditions. The enhanced chemical reaction rate could be due to the following reasons, as speculated by several researchers (Barnard et al. 2007; Kim et al. 2011; Leadbeater and Stencel 2006; Loupy et al. 1993; Vyas, Verma, and Subrahmanyam 2010).

- Energy transfer from microwaves to the material is believed to occur either through resonance or relaxation, which results in rapid heating and thus delivers energy directly to the reactant.
- Microwave assists more molecular friction and collisions in reaction medium, giving rise to intense localized heating and thereby accelerating the chemical reaction.

Other important areas are better fundamental understanding and modeling of microwave-material interactions, better preparation of reaction mixtures and compositions tailored specifically to microwave processing, better process controls, electronic tuning, and automation (smart processing). Finally, the availability of low-cost equipment, supporting technologies, and other processing support hardware is to be considered. Combining the microwave effect with other innovative heating methods can be beneficial. Ultrasonics and radiofrequency waves can complement the microwave effect to improve the overall reaction performance in hybrid reactors; the use of ultrasonic technology seems especially promising. Research in this area is in its infancy; however, if successfully demonstrated, a combined effect of these two innovative technologies can be enormous.

2.1.6 STATIC MIXER

Static mixers are simple devices consisting of spiral-shaped internal parts within an enclosure, such as a tube or pipe, that promote turbulent flow.

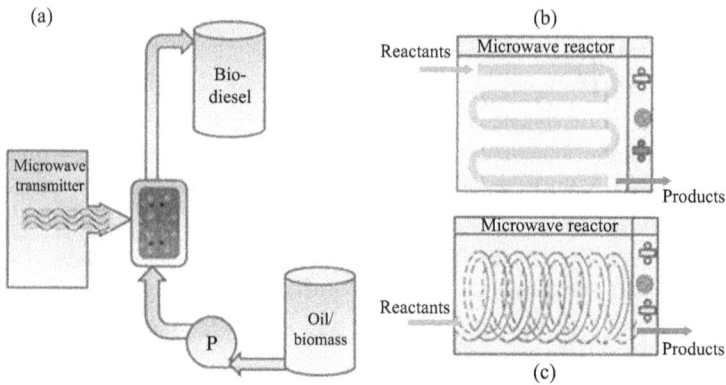

Figure 2.5. Microwave reactor for biodiesel production.

They have no moving parts, are easy to use and maintain, and are very effective at mixing liquids that are not readily miscible under normal conditions. The experiment was conducted by Thompson and He (2007) shown in Figure 2.6. The system is composed of two stainless steel static reactors (0.9 mm ID×300 mm long) including 34 fixed right- and left-hand helical mixing elements. According to Charles Ross & Son Company, the static mixture components inside the mixing chamber are subjected to high levels of mechanical and hydraulic shear as the rotor turns within a close tolerance stator at tip speeds ranging from 3,000 to 4,000 ft/min. The resulting mixture is then expelled at high velocity through holes in the stationary stator. A study was conducted to explore the possibility of using a static mixer as a continuous-flow reactor for biodiesel production (Thompson and He 2007). The results showed that the static mixer reactor was effective for biodiesel production, and products meeting the ASTM D6584 specification were obtained. As with other reactor configurations, temperature and catalyst concentration influenced the product yield significantly. The most favorable conditions for complete transesterification were 60°C and 1.5 percent catalyst for 30 min. It is feasible, therefore, to use a static mixer alone as the reactor for biodiesel preparation from vegetable oils and alcohols. A similar process is sometimes used commercially, but the use of a large static mixer as the biodiesel processor has not been commercialized.

2.1.7 SUPERCRITICAL REACTOR

An alternative method to avoid the catalyst requirement, transesterification can be achieved in a catalyst-free manner by using a "supercritical"

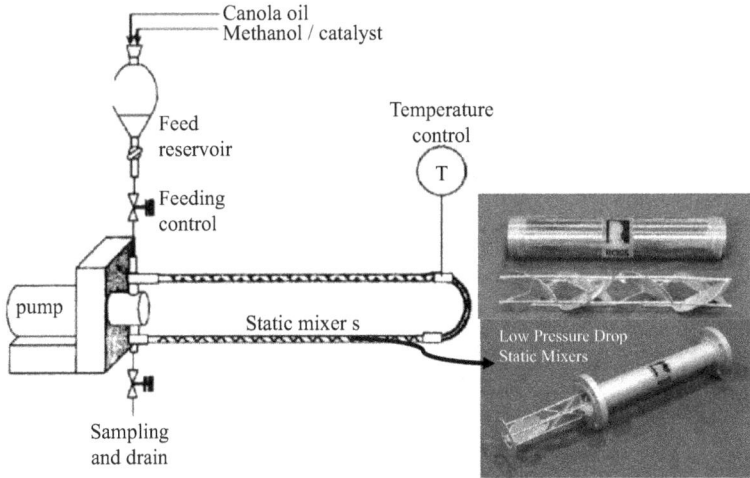

Figure 2.6. Experimental set-up of static mixer.

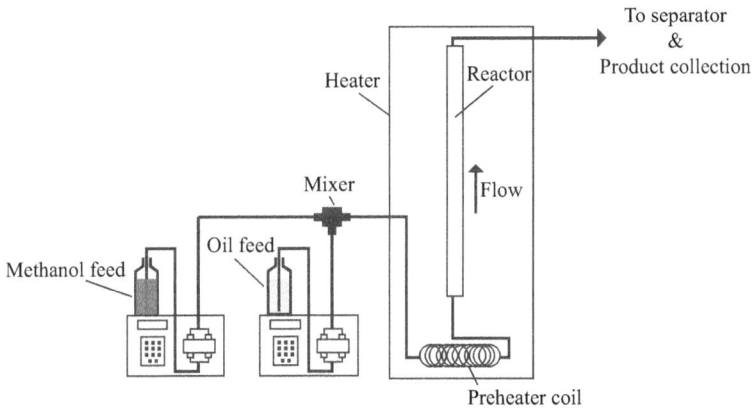

Figure 2.7. Biodiesel production using supercritical methanol.

process. A schematic diagram of a system for carrying out the reaction is shown in Figure 2.7. Work has so far been carried out in laboratories and on a pilot or small production scale. When transesterification occurs during the supercritical state of methanol (typically 300°C and 40 MPa/5800 psi or higher), the vegetable oil or animal fat dissolves in methanol to form a single phase. The reaction then occurs to reach completion in a few minutes without any catalysts. The supercritical process tolerates water and free fatty acids in the system, and the soap formation that is common in the traditional process is eliminated (Demirbas 2002, 2005, 2006; Saka

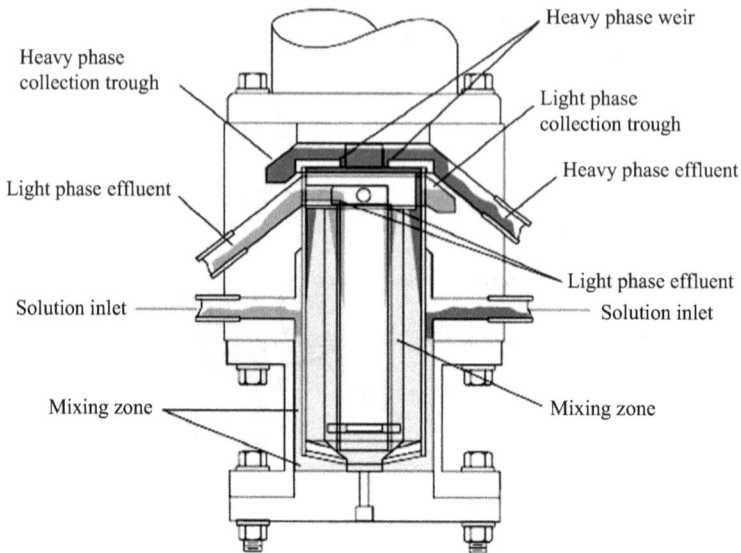

Figure 2.8. Centrifugal reactor for biodiesel production.

and Kusdiana 2001, 2004; Minami and Saka 2006; Shin et al. 2012). Since the supercritical state demands very high temperature and pressure, the process can be expensive. Nevertheless, large biodiesel producers may find this process to be cost effective because, since the reaction happens so quickly, producers can make a large quantity with a relatively small reactor and limited space.

2.1.8 CENTRIFUGAL CONTACT SEPARATOR

The method includes continuously contracting a triglyceride-containing component with an alcohol and a catalyst at an elevated temperature in a centrifugal reactor/separator. Centrifugal devices are widely used for separating materials of different densities. Such devices have been found to provide a highly satisfactory method of separating liquids from one another based on different weight phases. Recently, the use of the centrifugal reactor (Figure 2.8) for biodiesel synthesis was reported by Jennings (2008). According to Jennings (2008), the apparatus includes a stationary shell, a rotating hollow cylindrical component disposed in the stationary shell, a residence-time increasing device external to the stationary shell, a standpipe for introducing fluid into an interior cavity of the hollow cylindrical component from the residence-time increasing device, a first outlet in fluid flow communication with the interior cavity of the hollow cylindrical component for a less dense phase fluid, and a second outlet in fluid flow communication with the interior cavity of the hollow cylindrical component for a more dense phase fluid. This method is specifically applicable to the production of biodiesel through the esterification of organic oils and fats.

2.1.9 HIGH-FREQUENCY MAGNETIC IMPULSE CAVITATION REACTOR

The high-frequency magnetic impulse cavitation reactor is the third-generation hydrodynamic high-frequency magnetic-impulse reactor as reported by Gordon, Gorodnitsky, and Grichko (2013). Recently, the use of the high-frequency magnetic impulse cavitation reactor (Model: PULSAR-CT 215-B cavitation reactors as shown in Figure 2.9) for biodiesel synthesis was reported by Biofluidtech company. According to the company, the molecules of fatty acids are split with micro-explosions; this results in decrease of viscosity, increase of cetane number, improvement of energetic parameters of future fuel, as well as considerable acceleration and improvement of quality of esterification reaction. Moreover, the reaction goes by room temperature and there is no need to heat oil. It was highlighted that the traditional methods of biodiesel production are based on heating of oil up to 67°C to 70°C. It requires significant electric power inputs; besides the recovery of methanol (the necessary requirement for the reaction proceeding in traditional technologies) bring to great electric power consumption.

Figure 2.9. Magnetic impulse cavitation reactor developed by Biofluidtech.

With cavitation processing, there is no need of all the above-mentioned stages and the result is 5 to 7 times electric power saving.

2.1.10   MICRO-CHANNEL REACTOR

Microchannel reactors are compact reactors that have channels with diameters in the millimeter range. The small diameter channels dissipate heat more quickly than conventional reactors with larger channel diameters in the 2.5 to 10 cm (1 to 4 inch) range so more active catalysts can be used. The configuration of zigzag micro-channel reactors with narrower channel size developed by LeViness et al. (2011) is shown in Figure 2.10. It was reported by Šalić and Zelić (2011) that the microchannel reactors improve heat and mass transfer due to short diffusion distance and high volume/ surface area, so reaction rates achieved in them are rapid. Besides that, because of microreactor size, they offer reductions in construction and operating costs. Wen et al. (2009) investigated zigzag micro-channel reactors for continuous alkali-based biodiesel synthesis. At a residence time of 28 s and

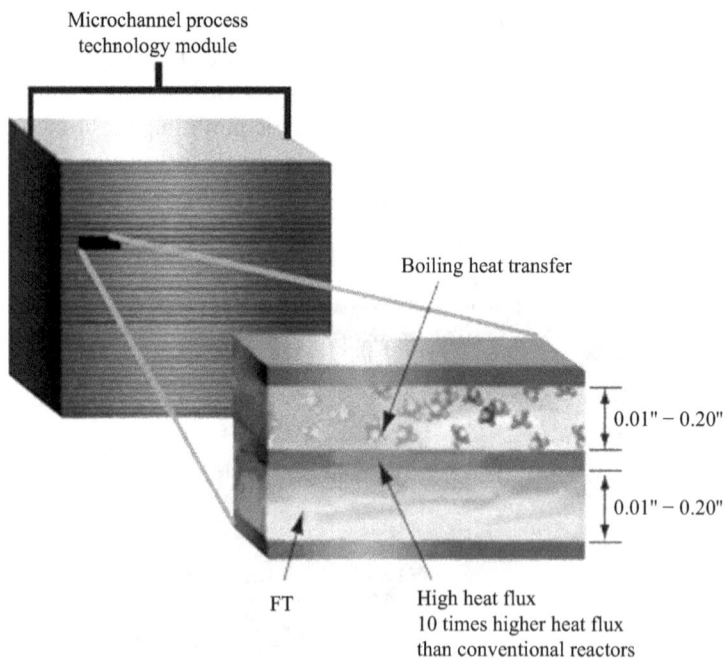

Figure 2.10. Microchannel reactor containing large numbers of parallel microchannel used for biodiesel production.

a temperature of 56°C, the yield of methyl ester reached 99.5 percent in an optimized zigzag micro-channel reactor using a 9:1 molar ratio of methanol to oil and a catalyst concentration of 1.2 wt% sodium hydroxide. Sun et al. (2008) studied KOH-catalyzed transesterification of unrefined rapeseed oil and cottonseed oil with methanol in capillary microreactors with inner diameters of 0.25 mm. At a 5.89 min residence time, they obtained a 99.4 percent yield of methyl esters at a catalyst concentration of 1 wt% KOH and using a 6:1 molar ratio of methanol to oil at a temperature of 60°C.

Despite the high percentage of fatty acid methyl ester (FAME) (>90 percent) achieved by powdered catalysts, many catalytic systems have not been commercialized because of the difficulties encountered when trying to separate such catalysts from the reaction media. In addition, the small particle size gives rise to several problems such as high pressure drops, poor mass/heat transfer, poor contact efficiency, and difficulties in handling and separation (Islam et al. 2013a; 2013b). Thus, this type is especially efficient for the use of solid catalyst for the production of biodiesel.

2.1.11   OSCILLATORY FLOW REACTORS

Flow reactors use a combination of flow oscillation and baffled tube geometry to ensure efficient mixing and effective heat transfer. A novel reactor system using oscillatory flow shown in Figure 2.11 was developed by Costello (2006). The standard reactor consists of an oscillator base and a reactor tube top section (Figure 2.11). A nutating cam mechanism driven by an electric motor and linear actuator controls the amplitude and frequency of operation. A pair of pistons driven off the two cams provides oscillations in an inverted U arrangement of reactor tubes. All of the variations are achieved by the electronic control of the motors. An oscillatory motion is superimposed upon the net flow of the process fluid, creating

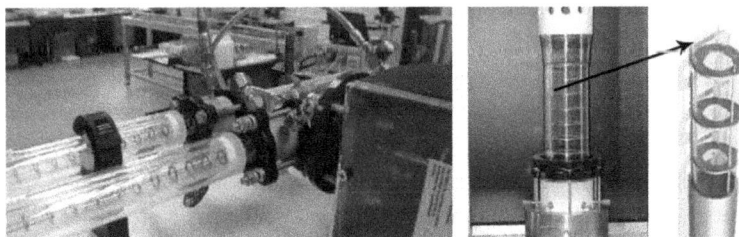

Figure 2.11. Standard design includes an oscillator base and a reactor tube top section.

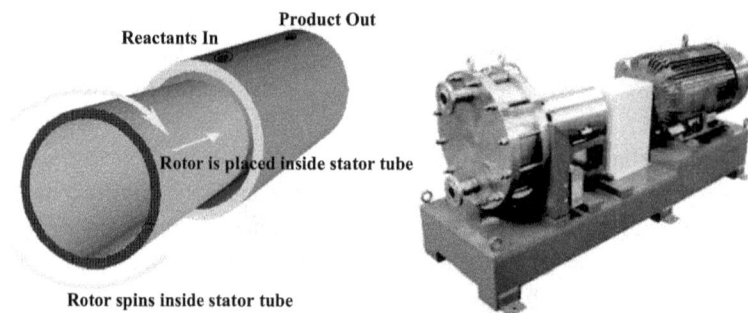

Figure 2.12. A commercial SPR reactor for biodiesel production developed by Hydro dynamics, Inc. (Courtesy of Hydro dynamics, Inc.).

flow patterns conducive to efficient heat and mass transfer, while maintaining plug flow (Phan, Harvey, and Rawcliffe 2011). Oscillatory motion in the tube is provided by an electrically or pneumatically driven piston or diaphragm to oscillate the fluid or to displace series of baffles (Mackley 1991). Many long residence time processes are currently performed in batch, as conventional designs of plug flow reactor prove to be impractical due to their high length-to-diameter ratios, which lead to problems such as high capital cost, large "footprint," high pumping costs, and also control is difficult. The oscillatory flow reactor (OFR) allows these processes to be converted to continuous, thereby intensifying the process, as reported by Harvey, Mackley, and Seliger (2003).

2.1.12 SPINNING DISC REACTOR

The spinning disc reactor (SDR) with a stator, which has been developed to improve the mixing and mass transfer between methanol and oil, thus increasing the product efficiency of biodiesel production. The rotating, or spinning, tube reactor is a shear reactor consisting of two tubes. A scheme of the rotating tube reactor developed by Hydro dynamics, Inc. is shown in Figure 2.12. The SDR includes a stationary disc, which is coaxially spaced adjacent to a rotating parallel disc separated only by a fraction of a millimeter. The configuration may lead to intense, forced molecular inter-diffusion of liquid–liquid two phases caused by high shear rate. Chen and Chen (2014) studied the biodiesel production using spinning disk reactor and concluded that the spinning disk reactor is a promising alternative method for continuous biodiesel production. The optimal yield of 96.9 percent was obtained with a residence time of 2 to 3 s at a molar ratio of 6, potassium hydroxide concentration of 1.5 wt%, temperature of 60°C, flow rate of 773 mL/min, and rotational speed of 2400 rpm. Qiu, Petera, and Weatherley (2012) reported that the residence time for the attainment of equilibrium was decreased 20- to 40-fold compared with that determined for a stirred batch reactor.

2.2 SEPARATION OF CRUDE BIODIESEL

The first step after transesterification is the separation of crude biodiesel from by-product, glycerol. With the separation between biodiesel and the by-product, glycerol is primarily achieved through different techniques as follows:

- Gravitational settling
- Centrifugation
- Filtration
- Decantation
- Sedimentation

The gravitational settling separation of biodiesel and glycerol is a result of differences in their polarities and also significant difference in their densities (Figure 2.13). As a rule of thumb, the difference in specific gravity of 0.1 in a mixture of compounds will result in phase separation by gravity. As can be seen in Table 2.1, gravity separation is suitable to recover biodiesel from the process byproducts (glycerine and methanol).

Biodiesel will be in a mixture of excess methanol, catalyst, and glycerine after the completion of the oil conversion reaction. However, impurities in the feedstock may cause emulsion formation, high interferences with phase separation. Saturated salt (sodium chloride) or centrifugation breaks the emulsion and speeds up the phase separation. Recently,

Table 2.1. Specific gravity of the compound

Compound	Specific gravity
Methanol	0.79
Biodiesel	0.88
Soybean oil	0.92
Catalyst	0.97
Glycerin	1.28

Figure 2.13. Gravitational separation of catalyst, glycerol, and biodiesel.

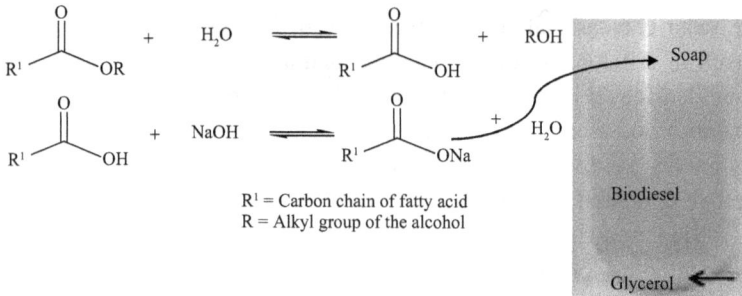

R^1 = Carbon chain of fatty acid
R = Alkyl group of the alcohol

Figure 2.14. Formation of soap and water.

Hayyan et al. (2010) have reported that the quaternary ammonium salt–glycerine-based ionic liquid can be effective for the separation of glycerine and other un-reacted reactants and by-products. A good conversion reaction will require excess methanol, but the amount of methanol in the system has to be minimized for good phase separation. The separation process between biodiesel and glycerol can be difficult in the presence of soaps formation (Figure 2.14), which mostly solidifies and forms a semi-solid substance (Van Gerpen 2005). The use of a large amount of demineralized water can be used to remove the residual soap from the biodiesel. The heterogeneous catalytic system could circumvent the problem associated with the homogeneous catalyst. Besides, the enzymatic alcoholysis could also alleviate separation difficulties commonly encountered with alkaline catalyst.

Gomes, Pereira, and Barros (2010) remarked that separation of biodiesel from glycerol via decantation is cost effective. However, the process requires a long period of time ranging from 1 to 8 hours to achieve good separation (Dubé, Tremblay,and Liu 2007). Therefore, to hasten the product separation process, centrifugation technique is mostly employed. The process of centrifugation is fast, but the cost involved is considerably high (Van Gerpen 2005).

2.3   METHANOL RECOVERY FROM BIODIESEL

Following transesterification, it is highly recommended that any remaining methanol be removed to avoid loss of potential reactant to the waste stream. Flash evaporation or boiling utilizes the low evaporation temperature of methanol to remove it from the FAME and methanol. Methanol acts as a cosolvent, which keeps some glycerin and soaps in solution with biodiesel. The methanol can then be recovered by condensing the vapor and reused for subsequent batches (Van Gerpen et al. 2006; Knothe and Steidley 2005). By separating glycerin prior to methanol recovery, the chance the reaction will reverse itself during the methanol recovery process will be diminished (Hatti-Kaul 2010).

2.4   REFINING OF BIODIESEL

The refining of crude biodiesel is usually achieved via two notable techniques: wet and dry washings. Conventionally, wet washing is the most employed technique to remove impurities such as soap, catalyst, glycerol, and residual alcohol from biodiesel. The refining of crude biodiesel is primarily done to achieve high purity and quality biodiesel products that can be used in compression–ignition (diesel) engines (Atadashi et al. 2012a). Furthermore, refining of crude biodiesel is a key factor to its commercial production and application. Thus, the continuous development of these refining technologies to purify biodiesel has raised hope for biodiesel industrial production and practical usability. Furthermore, achievement of high-quality biodiesel fuel could provide the following benefits: reduction in elastomeric seal failures; decrease in fuel injector blockages and corrosion due to absence of glycerol, catalysts, and soaps; reduction in degradation of engine oil thereby providing high engine performance; and better lubricant properties and better quality exhaust emissions (Fazal,

Haseeb, and Masjuki 2011). In addition, the generation of high-quality biodiesel could also lead to the elimination of fuel tank corrosion effects, eradication of bacterial growths and congestion of fuel lines and filters, and annihilation of pump seizures emanating owing to higher viscosity at low temperatures (Bunce et al. 2010; Snyder et al. 2010).

2.4.1   WET WASH

Biodiesel wet washing technique involves the addition of certain amount of water to crude biodiesel and agitating it gently to avoid the formation of emulsion. The process is repeated until colorless wash water is obtained, indicating complete removal of impurities.

Water wash: The application of distilled water coupled with gentle water washing eliminates the precipitation of saturated biodiesel and prevents the formation of emulsions (Figure 2.15). It was reported by Atadashi et al. (2011) that the distilled water washing at 50°C was most effective to purify the crude biodiesel. The water washing process reduced the methanol, free glycerol concentrations at the standard EN 14214 level.

Acid water wash: Acids such as phosphoric acid, sulfuric acid, and hydrochloric acid are mostly used in the purification of crude biodiesel.

Phosphoric acid: The 5 percent phosphoric acid with silica gel is the more suitable method for purification of crude biodiesel from frying oil (Stojković et al. 2014). The final biodiesels satisfied the criteria prescribed with the biodiesel fuel standards with respect to density, viscosity, acidity

Figure 2.15. Purification of crude biodiesel with distilled water (a) after vigorous mixing; (b) settling glycerol and biodiesel; and (c) after separation of glycerol.

(except that obtained from the waste cooking oil having the initial acid value of 3 mg KOH/g), iodine value, and purity.

Hydrochloric acid: The washing step with 5 percent hydrochloric acid water saturated with carbon dioxide can remove soap and glycerol completely from the crude biodiesel (Tan, Abdul Aziz, and Aroua 2013).

Sulfuric acid: Fatty acids in waste frying oils reacted with the alkali catalyst to form soap and water, which inhibit separation and purification of the biodiesel; thus, an esterification pretreatment with an acid catalyst (sulfuric acid, 98 percent) was conducted on waste frying oils to get rid of the inhibitory fatty acids (Jayasinghe, Sungwornpatansakul, and Yoshikawa 2014). It was reported by Li et al. (2008) that this pretreatment was necessary in order to easily and successfully purify the biodiesel.

Glycerol washing: Glycerol was the only solvent that positively affected the water content in the biodiesel. Soap and methanol were removed with glycerol washing to the same extent as with water washing. However, it influenced negatively to the free glycerol content. Di Felice et al. (2008) found that when up to 6 percent soap content by weight was added to the biodiesel, methanol, and glycerol mixture, almost all (97 to 99.55 percent) of the soap was found in the heavy glycerol-rich phase. It is reasonable to state that the soap can drag a large amount of biodiesel into either the glycerol or water phase and cause a loss of product. Jaruwat, Kongjao and Hunsom (2010) also recorded 6 to 7% v/v biodiesel and oil in the water phase.

Organic solvents wash: Petroleum ether and n-hexane is used to remove the residual alcohol from the biodiesel. The purification process losses the yield of biodiesel due to the formation of emulsion. It was reported by Kocherginsky, Yang, and Seelam (2007) that the use of hollow fiber membrane prevented the emulsification and reduced the purification losses.

2.4.2   DRY WASH

The dry washing technique involves the use of ion exchange resins and magnesol powder, activated fibers, activated carbon, activated clay, and acid clay to substitute water washing to remove biodiesel contaminants. A schematic diagram of biodiesel dry washing process was shown in Figure 2.16. This technique is also employed in commercial plants to purify biodiesel (Faccini et al. 2011).

Ion exchange resin: The ion exchange resin can reduce the glycerol content and soap from the crude biodiesel to the EN14214 standard specification. However, it has less effect on the removal of methanol.

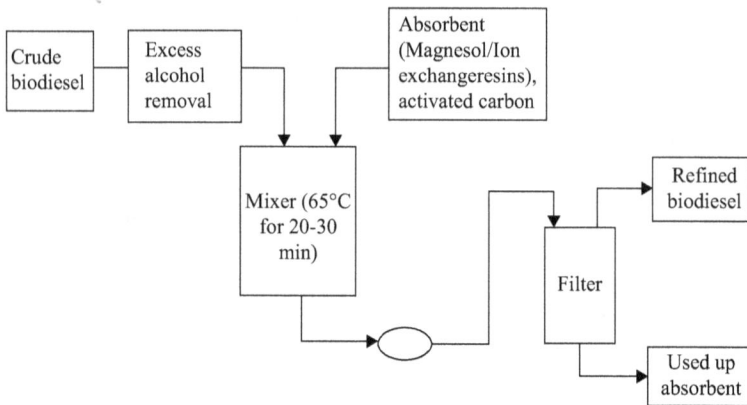

Figure 2.16. Schematic diagram of biodiesel dry washing process.

Magnesol powder: The inorganic matrix magnesol is a synthetic adsorbent composed of magnesium silicate and anhydrous sodium sulfate. Magnesol has a strong affinity for polar compounds, thereby actively filtering out metal contaminants, mono and di-glycerides, free glycerin, excess methanol, as well as free fatty acids and soap (D'ippolito et al. 2007; Lou, Zong, and Duan 2008; Faccini et al. 2011; Berrios et al. 2008). Yori et al. (2007) and Predojević (2008) studied the removal of glycerol from biodiesel from waste frying oils with elevated acid values using silica and achieved high purity of the resultant biodiesel. Therefore, during the washing process, the amount of glycerides and total glycerol in crude biodiesel are lowered to a reasonable level.

Amberlite BD10 DRY, Purolite (PD206) resin: Purolite (PD206) is a dry polishing media specifically formulated to remove maximum amounts of waste water, soaps, salts, catalysts, and glycerol during the biodiesel purification process (Banga, Varshney, and Kumar 2012). The organic resin Amberlite BD10 DRY is already being used in pilot industries, where biodiesel is purified through a column filled with the resin (Faccini et al. 2011). Both resins were also studied by Berrios and Skelton (2008) who investigated these ion exchange resins by passing the feed through a column of resin supported in a glass tube at room temperature.

Activated fibers, activated carbon, activated clay, and acid clay: Biodiesel washing with clay, especially acid clay treated with sulfuric acid, is preferable, which is superior in the aspects of de-alkaline effect, deodorant effect, and decoloring effect. Also, clay grain size ranging from 0.1 mm to 1.5 mm is more suitable for effective biodiesel purification. Atabani et al. (2013) stated that clay with smaller grain size provides superior purification process, but separation after the purification treatment is more difficult. However, when the clay grain size is larger, separation after the treatment becomes easier, but the purification process is inferior.

A wet washing process usually requires a lot of water, approximately water wash solution at the rate of 28 percent by volume of oil and 1 g of tannic acid per liter of water. The use of large quantity of water generates a huge amount of wastewater and incurs high energy cost. Thus for each liter of biodiesel, close to 10 L of wastewater is produced. Refining of crude biodiesel alone accounts for 60 to 80 percent of the total processing cost. Dry washing appears to be a promising method for refining crude biodiesel, but the disposal of a large amount of solid waste generated from the dry washing technique is the main hurdle. Thus, more efforts need to be made to explore and exploit better purification processes such as membranes to effectively replace conventional biodiesel separation and washing techniques.

2.5   MEMBRANE TECHNOLOGY

A membrane reactor is a device for simultaneously carrying out a reaction and membrane-based separation in the same physical enclosure. A schematic diagram of the biodiesel membrane reactor system is shown in Figure 2.17. The idea of using membrane reactor for biodiesel production was first proposed by Dubé, Tremblay, and Liu (2007) and ever since quantitative researches were largely conducted. Due to the immiscibility of lipid feedstock and alcohol, lipids form droplets that are excluded from passing through the membrane pores. The micro-porous inorganic membrane selectively permeates FAAE, alcohol, and glycerol while retaining the emulsified oil droplets. Cao, Dubé and Tremblay (2008), Gomes, Pereira, and Barros (2010), Gomes, Arroyo, and Pereira (2011), Jiang et al. (2009), and Shuit et al. (2012) have reported on the production of biodiesel using the membrane reactor. Dubé, Tremblay, and Liu (2007) developed the two-phase membrane reactor to develop a continuous reaction process for the production of biodiesel and overcome the challenges due to mass transfer limitations, incomplete conversion, use of high FFA feedstock, and downstream purification. The experiments were performed in the membrane reactor in semi-batch mode at 60°C, 65°C, and 70°C and at different catalyst concentrations and feed flow rates. A conclusion has been reached by Dubé, Tremblay, and Liu (2007) that the reactor enabled the separation of reaction products (FAME/glycerol in methanol) from the original canola oil feed. Moreover, the two-phase membrane reactor was particularly useful in removing unreacted canola oil from the FAME product yielding high purity biodiesel and shifting the reaction equilibrium to the product side, as reported by Dubé, Tremblay, and Liu (2007).

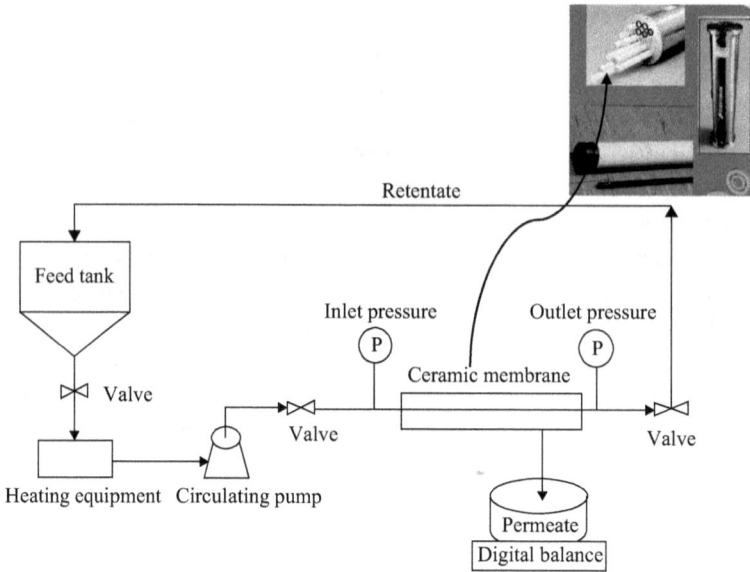

Figure 2.17. Experimental setup of biodiesel membrane separation.

Water washing has been the most frequently used process for purification of crude biodiesel, although it suffered from several drawbacks including the water cost, possible emulsion formation, drying of biodiesel, and wastewater treatment. Dry washing appears to be a promising method for refining crude biodiesel, but a solution for the generated solid waste should be found. Membrane technology circumvents the need for traditional water washing of crude biodiesel and might be a future choice for crude biodiesel refining. The membrane extraction results in avoiding the emulsification that is characteristic for water washing and in reducing the environmental problem of disposal (little or no wastewater generation), but it increases the final biodiesel production cost. Further research should be performed in order to make use of ionic liquids for crude biodiesel refining. The main challenges are to reduce their production cost, to develop easy methods for their recovery, and to develop effective methods for their recycling. Also, a large portion of membrane separation processes are carried out under moderate temperature and pressure conditions and their scale-up are less cumbersome. Furthermore, membranes are generally most preferred in the refining processes for the following reasons: low energy consumption; safety; simple operation; elimination of wastewater treatment; easy change of scale; higher mechanical, thermal, and chemical stability; and resistance to corrosion.

CHAPTER 3

BIODIESEL PRODUCTION PROCESSES

Several processes for biodiesel production have been developed (Figure 3.1). At present, the majority of biodiesel plants are operated either in batch or continuous mode using conventional homogeneous acid or alkali-based transesterification conversion technology. In homogeneous catalysis, alkali-catalyst is much suited for the transesterification of vegetable oils because the process proceeds much rapid than the acid-catalyzed reaction whereby acid-catalyst is usually used for the esterification of free fatty acid (FFA). Due to the fact that the alkali-catalysts are less corrosive than acid catalysts, alkali-catalysts such as sodium hydroxides (NaOH), sodium methylate, alkali metal alkoxides, sodium, or potassium carbonates are usually a preferred choice in industrial processes.

3.1 HOMOGENEOUS BASE-CATALYZED PROCESS

The base-catalyzed transesterification of vegetable oils proceeds faster than the acid-catalyzed reaction. During the transesterification reaction, the triglyceride is converted stepwise to diglyceride and monoglyceride intermediates, and finally to glycerol (Demirbas 2008a; Demirbas 2008b; Islam et al. 2013a; Taft, Newman, and Verhoek 1950). The mechanism of the base-catalyzed transesterification of vegetable oils is shown in Figure 3.2. The first step is the reaction of the base with the alcohol, producing an alkoxide and the protonated catalyst. The nucleophilic attack of the alkoxide at the carbonyl group of the triglyceride generates a tetrahedral intermediate (step 2) from which the alkyl ester and the corresponding anion of the diglyceride are formed (step 3). The latter deprotonates the catalyst, thus regenerating the active species (step 4), which is now able

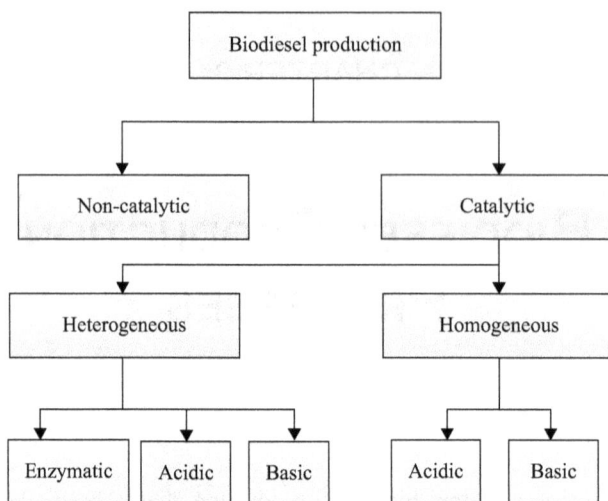

Figure 3.1. Biodiesel production process.

to react with a second molecule of the alcohol, starting another catalytic cycle. Diglycerides and monoglycerides are converted by the same mechanism to a mixture of alkyl esters and glycerol.

Depending on the quality of the feedstock, either esterification or transesterification reactions are used for biodiesel production. Most of the current biodiesel production operations use base catalysis (transesterification). Figure 3.3 shows a schematic diagram of the unit operations involved in base-catalyzed biodiesel production. This method works well if the FFA, moisture, and phosphorous contents of oil/fat are less than 0.1 percent, and less than 10 ppm, respectively (Dunford 2007). Typical feedstock for biodiesel production are soybean, canola/rapeseed, sunflower, cottonseed, palm seed and palm kernel, corn, and mustard seed oil. Pork, beef, poultry fat, and grease can also be converted to biodiesel. Palm oil and animal fat may have a high FFA content, which causes soap formation that has adverse effects on downstream processing and leads to yield reduction. Sodium hydroxide (NaOH), potassium hydroxide (KOH), and sodium methoxide (CH_3ONa) are the most common catalysts for transesterification. Sodium methylate (sodium methoxide) is more effective than NaOH and KOH as a catalyst, but it is more expensive. Sodium methoxide is sold as a 30 percent solution in methanol for easier handling. Base catalysts are very sensitive to the presence of water and FFAs. The amount of sodium methoxide required is 0.3 to 0.5 percent of the weight of the oil. A higher amount of catalyst

Step 1 $ROH + B^- \rightleftharpoons RO^- + BH$

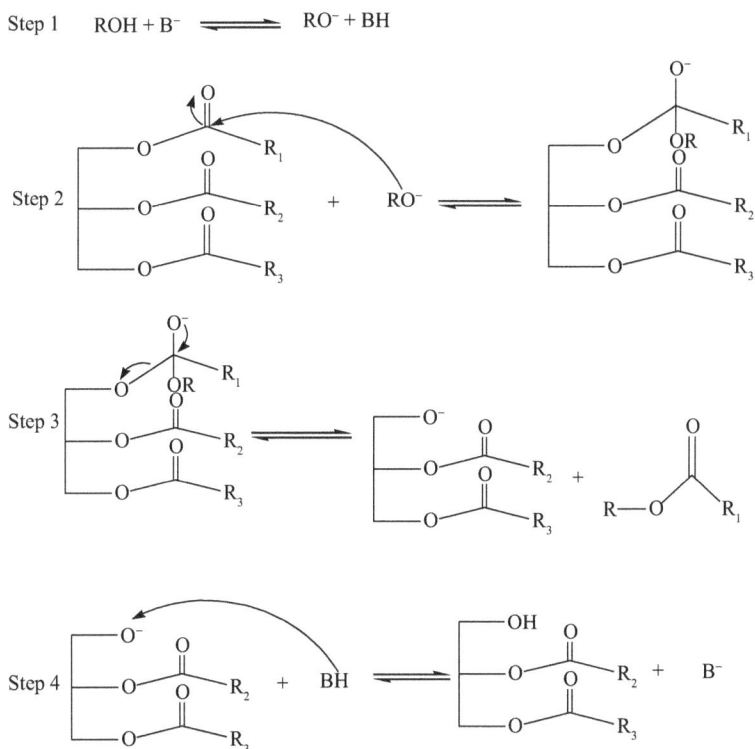

Figure 3.2. Homogeneous base-catalyzed mechanism for the transesterification of triacylglycerides.

(0.5 to 1.5 percent of the weight of the oil) is required when NaOH or KOH is used. NaOH and KOH also lead to water formation, which slows the reaction rate and causes soap formation (Demirbas 2008b; Sharma, Singh, and Upadhyay 2008; Singh et al. 2006; Zhang et al. 2003a; Zhang et al. 2003b). Methanol is the most common alcohol used for conversion of fats and oils to biodiesel. Methanol is flammable, so proper handling is required for safety. Transesterification is a reversible reaction. Thus, excess methanol is required to shift the equilibrium favor ably. It was reported (Dunford 2007) that theoretical biodiesel yields for biodiesel soybean and tallow were as follows: 1004.2 kg/1000 kg soybean and 998.1 kg/1000 kg tallow.

Methanol and oil do not mix well, and poor contact between the oil and methanol reactants means the reaction rate is slow. Vigorous mixing at the beginning of the reaction improves reaction rates. Near the end of the reaction, reduced mixing helps the separation of glycerine, and the

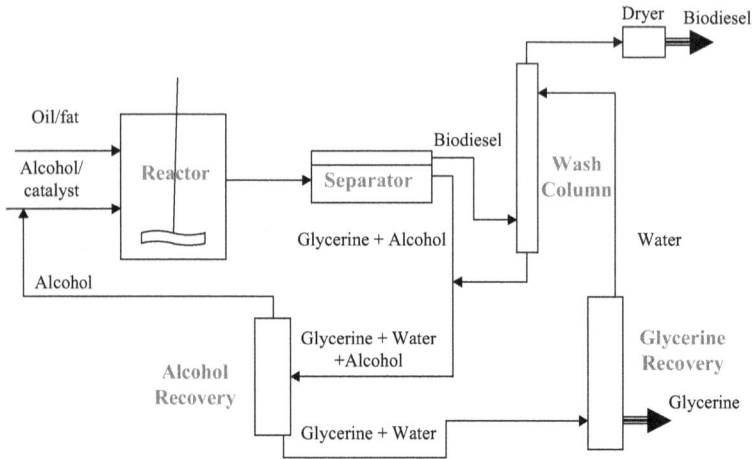

Figure 3.3. The flow diagram for the homogeneous acid or base catalyzed transesterification (Dunford 2007).

reaction would proceed faster in the top layer, which is oil and methanol. Odin et al. reported that the base reaction takes 4 to 8 hrs to complete at an ambient temperature (21°C). The reaction is usually conducted below the boiling point of methanol (60°C). At this temperature, the reaction time may vary between 20 min to 1.5 hrs. A higher temperature will decrease reaction times, but this requires the use of a pressure vessel because the boiling point of methanol is 65°C (Dunford 2007). The reactor is either sealed or equipped with a condenser to minimize alcohol evaporation during the conversion process. Higher oil conversion rates can be achieved if the production system is set up as a two-step process with two reactors. In such cases, glycerine formed in the first reactor is removed, and the reaction is completed in the second reactor.

3.2   HOMOGENEOUS ACID-CATALYZED PROCESS

The transesterification process is catalyzed by BrØnsted acids, preferably by sulfonic and sulfuric acids (Freedman, Butterfield, and Pryde 1986; Freedman, Kwolek, and Pryde 1986; Stern and Hillion 1990). These catalysts give very high yields in alkyl esters, but the reactions are slow, requiring, typically, temperatures above 100°C and more than 3 hrs to reach complete conversion (Freedman, Pryde, and Mounts 1984). Pryde (1983) reported that the methanolysis of soybean oil, in the presence of

1 mol% of H_2SO_4 with an methanol/oil molar ratio of 30:1 at 65°C, takes 50 hrs to reach complete conversion of the vegetable oil (>99 percent), while the butanolysis (at 117°C) and ethanolysis (at 78°C), using the same quantities of catalyst and alcohol, take 3 and 18 hrs, respectively.

The transesterification chemical pathway shown in Figure 3.4, for an acid-catalyzed reaction, indicates how in the catalyst-substrate interaction the key step is the protonation of the carbonyl oxygen (Demirbas 2008b). This in turn increases the electrophilicity of the adjoining carbon atom, making it more susceptible to nucleophilic attack (Meher, Vidya Sagar, and Naik 2006). In contrast, base catalysis takes on a more direct route, creating first an alkoxide ion, which directly acts as a strong nucleophile, giving rise to a different chemical pathway for the reaction (Figure 3.2). This crucial difference, that is, the formation of a more electrophilic species (acid catalysis) versus that of a stronger

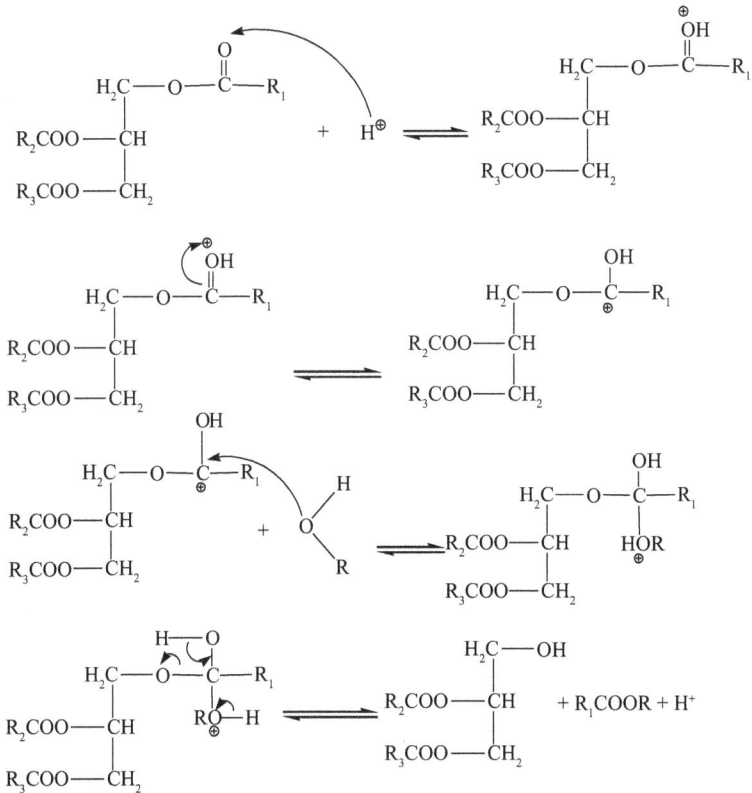

Figure 3.4. Homogeneous acid-catalyzed mechanism for the transesterification of triacylglycerides.

nucleophile (base catalysis), is ultimately responsible for the observed differences in activity.

Dunford (2007) reported that the base catalysis is not suitable if the FFA content of the feedstock is greater than 1 percent. There are two approaches for handling high FFA content feedstock. One way would be to refine the feedstock before base catalysis. FFAs can be removed by chemical neutralization or physical deacidification. Chemical neutralization involves treatment with caustic NaOH or KOH. Soap formed during this process is removed, and the remaining oil is ready for base catalysis. However, some oil is lost during this process. Physical deification, or steam stripping, also removes FFAs. This process is performed under vacuum and requires steam. Fats and oils with high FFA content can be converted to biodiesel using acid catalysis, which is the second approach for handling high FFA content feedstock (Dunford 2007). This technique uses a strong acid. Soap formation is not a problem because there are no alkali metals in the reaction medium. Acid catalysts can be used for transesterification of the triglycerides, but the reaction might take several days to complete (Di Serio et al. 2005; Lotero et al. 2005; Soriano, Venditti, and Argyropoulos 2009). This is too slow for industrial processing. Acid catalysis is also used for direct esterification of oils with high FFA content or for making esters from soap stock, which is a byproduct of edible oil refining. The esterification of FFAs to alcohol esters is relatively fast; it would take about 1 hr at 60°C to complete the reaction.

Water is formed during this reaction. To improve reaction rates, water needs to be removed from the reaction medium by phase separation. Acid catalysis requires a high alcohol to FFA ratio (20:1 or 40:1 mole ratio) and a large amount of catalyst (5 to 25 percent) (Lam and Lee 2010; Lam, Lee, and Mohamed 2010). Sulfuric acid and phosphoric acid are the most common acid catalysts. The feedstock is sometimes dried to 0.4 percent water and filtered before the reaction. Then, an acid and methanol mixture is added to the feedstock. Once the conversion of the fatty acids to methyl esters has reached equilibrium, the methanol, water, and acid mixture is removed by settling or centrifugation. Fresh methanol and base catalyst are added into the remaining oil for transesterification. Reaction times of 10 min to 2 hrs have been reported. Both transesterification and esterification reactions can be operated either as a batch or continuous process. A batch process is better suited to smaller plants that produce less than 1 million gallons per year and provide operation flexibility (Obaja et al. 2003). Continuous processing allows the use of high-volume separation systems, and therefore increases throughput.

3.3 VARIABLES AFFECTING IN HOMOGENEOUS ACID/BASE TRANSESTERIFICATION

The process of transesterification is affected by various factors depending upon the reaction condition used. The effects of these factors are described in the following text.

3.3.1 EFFECT OF FFA AND MOISTURE

The FFA and moisture content are key parameters for determining the viability of the vegetable oil transesterification process. The higher the acidity of the oil, smaller is the conversion efficiency. Bojan and Durairaj (2012) investigated the biodiesel production from *Jatropha curcas* oil containing high FFA. The presence of high FFA concentration (8.67 percent) was reduced the biodiesel yield significantly (80.5 percent) in one step conventional base catalyzed transesterification. Therefore, the author suggested to use a two-step acid pretreatment esterification and base-catalyzed transesterification process to improve the yield. It was reported that the FFA concentration of *J curcas* oil was reduced to 1.12 percent during the first step and in the second step, alkali-based transesterification gave 93 percent yield (Bojan and Durairaj 2012). It was reported by Narasimharao, Lee, and Wilson (2007) that the addition of more sodium hydroxide catalyst compensates for higher acidity, but the resulting soap consumes the catalyst and reduces the catalytic efficiency, as well as causing an increase in viscosity, the formation of gels, and difficulty in achieving separation of glycerol. These high FFA content oils/fats are processed with an immiscible basic glycerol phase so as to neutralize the FFAs and cause them to pass over into the glycerol phase by means of monovalent alcohols.

Turck (2003) investigated the influence of base-catalyzed transesterification of triglycerides containing substantial amount of FFA. It was recommended by Narasimharao, Lee, and Wilson (2007) that the FFA value lower than 3 percent is needed for the completion of base-catalyzed transesterification reaction. The prolonged contact with air will diminish the effectiveness of these catalysts through interaction with moisture and carbon dioxide. Thus, it was suggested that the methoxide and hydroxide of sodium or potassium should be maintained in anhydrous state. Besides, the saponification reaction can be controlled by maintaining the anhydrous state of glycerides and alcohol, as reported by Elst, Sijben, and Van Ginneken (2011).

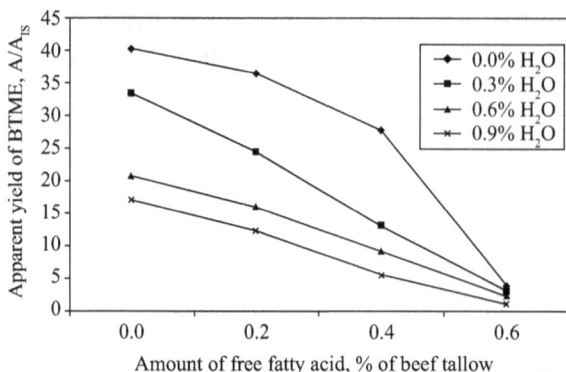

Figure 3.5. Effect of FFAs and water on the transesterification.

Ma, Clements, and Hanna (1998) studied the effect of FFA and water on the transesterification on beef tallow (Figure 3.5). Without adding FFA and water, the apparent yield of beef tallow methyl esters (BTME) was highest. When 0.6 percent of FFA was added, the apparent yield of BTME reached the lowest, less than 5 percent, with any level of water added. When 0.9 percent of water was added, without addition of FFA, the apparent yield was about 17 percent. If the low qualities of beef tallow or vegetable oil with high FFA are used to make biodiesel fuel, they must be refined by saponification using NaOH solution to remove FFAs. Conversely, the acid-catalyzed process can also be used for the esterification of these FFAs.

The problem with processing these low cost oils and fats is that they often contain large amounts of FFAs that cannot be converted to biodiesel using alkaline catalyst. Therefore, the two-step esterification process is required for these feed stocks. Initially, the FFA of these can be converted to fatty acid methyl esters (FAMEs) by an acid-catalyzed pretreatment and in the second step transesterification is completed by using alkaline catalyst to complete the reaction. Initial process development was performed with a synthetic mixture containing 20 and 40 percent FFA prepared by using palmitic acid. Process parameters such as molar ratio of alcohol to oil, type of alcohol, amount of acid catalyst, reaction time, and FFA level were investigated to determine the best strategy for converting the FFAs to usable esters. Canakci and Van Gerpen (2003) reported that the acid level of the high FFAs feed stocks could be reduced to less than 1 percent with a two-step pretreatment reaction.

Recently, the American researched-based learning network (eXtension) reported an alternative process called "glycerolysis" that can be used with feedstock containing more than 10 percent FFAs. According to the process condition, addition of glycerin at 400°F and letting it react with the FFAs to form monoglycerides, a glycerol molecule to which one FFA has been joined. These monoglycerides can then be processed using a standard alkaline catalyst transesterification process. Waste glycerin from biodiesel processing can be used in this process. Glycerolysis can be expensive because of the high heat involved, which requires a high-pressure boiler and trained boiler operator. Also, vacuum must be applied while heating to remove water that is formed during the reaction. Another disadvantage is that the glycerin will also react with the triglycerides in the oil to convert some of them to monoglycerides. While this does not negatively impact the reaction, it means that more glycerin is required for the process, and therefore more glycerin must be removed at the end of the transesterification.

Otherwise, the reaction should be performed under supercritical conditions (275 to 325°C and high pressure). At high temperature and pressure, the reaction does not require a catalyst, so soap formation is not a problem. Water also does not appear to inhibit the reaction. Both FFAs and triglycerides react easily, so there is no need to separate these materials before processing. In fact, even very low quality feedstock can be processed successfully. However, the high reaction pressure requires heavy-duty reaction vessels. The reaction conditions are so extreme that many side reactions can occur, which produce undesired products. The formation of these non-ester compounds means that the final product will probably need to be distilled to meet the American Society for Testing and Materials (ASTM) quality requirements. In addition, another drawback is the extra energy needed to achieve and maintain the high temperature. In spite of these concerns, this method is of interest because it allows processors to make use of low-cost feedstock such as trap grease.

3.3.2 TYPE OF CATALYST AND ITS CONCENTRATION

A catalyst functions to accelerate the reaction rates. For transesterification reaction, an increasing amount of heterogeneous catalyst caused the slurry (mixture of catalyst to reactant) to be too viscous giving rise to a problem of mixing and a demand of higher power consumption for adequate stirring. On the other hand, when the catalyst loading amount was not enough, maximum production yield could not be reached. To avoid this kind of problem, an optimum amount of catalyst concentration had to be investigated. Homogeneous catalysts used for the transesterification of triglycerides are classified as alkali or acid. The most notable catalyst used in producing biodiesel is the homogeneous alkaline catalyst such as NaOH, KOH, CH_3ONa, and CH_3OK. The choice of these catalysts is due to their higher kinetic reaction rates. Freedman, Kwolek, and Pryde (1986) showed that $NaOCH_3$ is a more effective catalyst formulation than NaOH and almost equal oil conversion was observed at 6:1 alcohol-to-oil molar ratio for 1%wt NaOH and 0.5%wt $NaOCH_3$. Methanolysis of beef tallow was studied by Ma, Clements, and Hanna (1998) with catalysts NaOH and NaOMe. Comparing the two catalysts, NaOH was significantly better than NaOMe. The catalysts NaOH and NaOMe reached their maximum activity at 0.3 and 0.5% w/w of the beef tallow, respectively. Vicente, Martınez, and Aracil (2004) reported higher yields with methoxide catalysts, but the rate of reaction was highest for NaOH and lowest for $KOCH_3$ at 65°C, a methanol-to-oil ratio of 6:1, and a catalyst concentration of 1%wt.

If the oil has high FFA content and more water, acid-catalyzed transesterification is suitable. The acids could be sulfuric acid, phosphoric acid, hydrochloric acid, or organic sulfonic acid. The acid-catalyzed transesterification was studied with waste vegetable oil (Canakci and Van Gerpen 1999) and showed that the same concentration of HCl and H_2SO_4 in the presence of 100 percent excess alcohol decreases the viscosity. H_2SO_4 has superior catalytic activity in the range of 1.5 to 2.25 M concentration (Fukuda, Kondo, and Noda 2001). Although chemical transesterification using an alkaline catalysis process gives high conversion levels of triglycerides to their corresponding methyl esters in short reaction times, the reaction has several drawbacks: It is energy intensive, recovery of glycerol is difficult, the acidic or alkaline catalyst has to be removed from the product, alkaline waste water requires treatment, and FFA and water interfere in the reaction.

However, because of high cost of refined feedstock and difficulties associated with the use of homogeneous alkaline catalysts to transesterify low quality feedstock for biodiesel production, the development of various heterogeneous catalysts are now on the increase. A new class of heterogeneous biodiesel catalyst was developed that has the ability to convert less refined and less costly oil feedstock to biodiesel while, at the same time, simplify the biodiesel production process as declared inventor company (NextCAT 2014). The commercializing biodiesel catalyst technology (NextCAT) has recently announced a new generation of catalysts that could convert oil feedstock with FFA as high as 30 percent and water up to 5 percent with a >98 percent FAME yield (Figure 3.6). In addition, these same catalysts have shown the capability to simultaneously process both FFA, and mono, di, and triglycerides, which allows

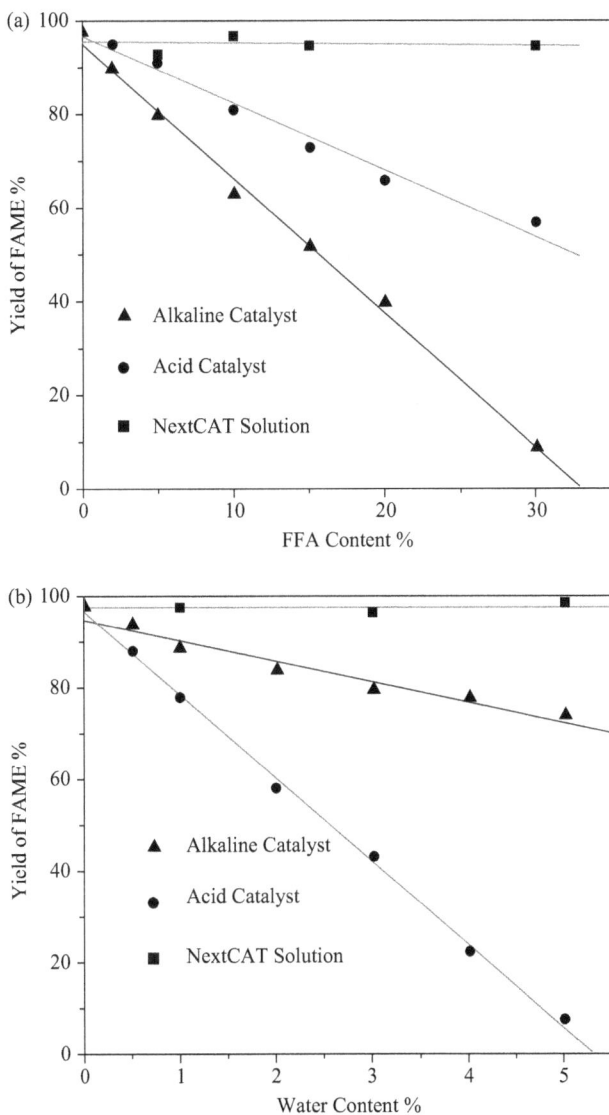

Figure 3.6. Effect of (a) FFA content and (b) water of the catalyst developed by NextCAT.

biodiesel producers to use less refined and less costly feedstock such as residual corn oil, waste vegetable oil, and brown grease. This greatly simplifies the FAME/glycerin separation and drying processes, and eliminates much of the current disposal costs associated with soap extraction and hazardous by-products.

3.3.3 MOLAR RATIO OF ALCOHOL TO OIL AND TYPE OF ALCOHOL

Another important variable affecting the ester yield is the molar ratio of alcohol to vegetable oil. Stoichiometrically, the methanolysis of vegetable oil requires three moles of methanol for each mole of oil. Since trans-esterification of triglycerides is a reversible reaction, excess methanol is required to shift the equilibrium toward the direction of ester forma-tion. The transesterification of groundnut oil with ethanol was studied by Oghome (2012) at molar ratios between 6:1 and 15:1 (Figure 3.7). As the mote ratio was increased to 6:1 and 9:1, there was a drop in ester yield. For a molar ratio of 12:1, the separation of glycerin is difficult and the apparent yield of esters decreased because a part of the glycerol remains in the biodiesel phase. Therefore, molar ratio 9:1 seems to be the most appro-priate. Furthermore, Kumar et al. (2010b) have obtained above 98 percent yield using 1:9 *Jatropha* oil to methanol molar ratio and the heterogeneous solid catalyst used was $NaSiO_2$. No methyl ester yield was achieved using a 4:1 molar ratio, and this can be attributed to the predominance of ester-ification reaction at the initial phase of transesterification to transesterify the FFA present in the *Jatropha* oil, which can consume methanol present in the reaction mixture, and, hence, the amount of methanol available for transesterification may not be sufficient to drive the reaction forward for longer time.

Krisnangkura and Simamaharnnop (1992) studied the continuous transesterification of palm oil at 70°C using organic solvent with sodium

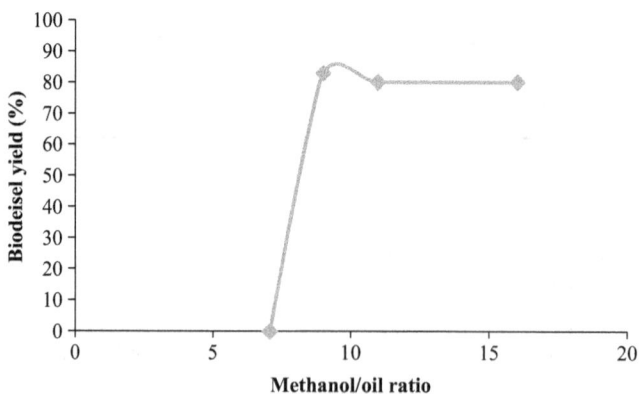

Figure 3.7. Effect of mole ratio of methanol to oil on biodiesel yield at constant temperature of 70°C, time 20 min, and catalyst concentration 0.5% w/w.

methoxide as a catalyst and found that the conversion increased with increasing molar ratios of methanol to palm oil. Thus, a molar ration of 6:1 is normally used in industrial processes to obtain methyl ester yields higher than 98 percent on a weight basis. Freedman, Pryde, and Mounts (1984) and Freedman (1986) studied the effect of molar ratios (from 1:1 to 6:1) on ester conversion with vegetable oils. Soybean, sunflower, peanut, and cotton seed oils behaved similarly, with the highest conversion being achieved at a 6:1 molar ratio.

The base-catalyzed formation of ethyl ester is difficult compared to the formation of methyl esters. The effect of alcohol on the transesterification reaction is shown in Figure 3.8. Specifically the formation of stable

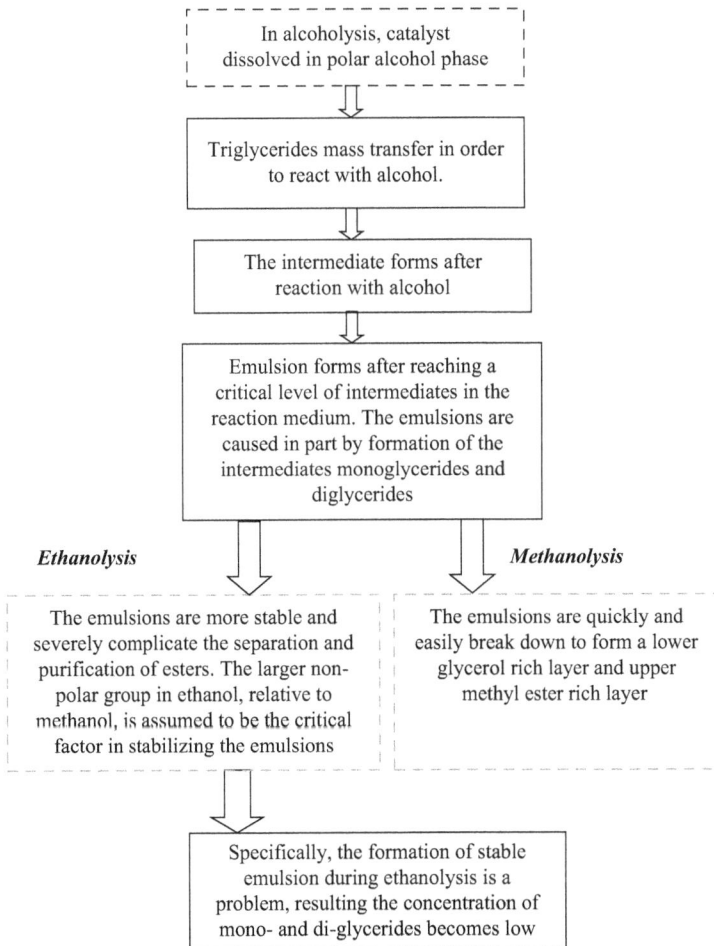

Figure 3.8. The effect of alcohol in the transesterification reaction.

emulsion during ethanolysis is a problem. Methanol and ethanol are not miscible with triglycerides at ambient temperature, and the reaction mixtures are usually mechanically stirred to enhance mass transfer. During the course of reaction, emulsions usually form. In the case of methanolysis, these emulsions quickly and easily break down to form a lower glycerol-rich layer and upper methyl ester rich layer. In ethanolysis, these emulsions are more stable and severely complicate the separation and purification of esters (Tomasevic and Siler-Marinkovic 2003). The molar ratio has no effect on acid, peroxide, saponification, and iodine value of methyl esters (Zhou, Konar, and Boocock 2003). However, the high molar ratio of alcohol to vegetable oil interferes with the separation of glycerin because there is an increase in solubility. When glycerin remains in solution, it helps drive the equilibrium back to the left, lowering the yield of esters.

3.3.4 EFFECT OF REACTION TIME AND TEMPERATURE

The conversion rate increases with reaction time. Freedman et al. (1986) observed the increase in fatty acid esters conversion when there is an increase in reaction time. The reaction is slow at the beginning due to mixing and dispersion of alcohol and oil. After that the reaction proceeds very fast. However, the maximum ester conversion was achieved within <90 min. Further increase in reaction time does not increase the yield product, that is, biodiesel/mono alkyl ester (Leung and Guo 2006; Alamu et al. 2007). Besides, longer reaction time leads to the reduction of end product (biodiesel) due to the reversible reaction of transesterification resulting in loss of esters as well as soap formation (Eevera, Rajendran, and Saradha 2009). Ma, Clements, and Hanna (1998) studied the effect of reaction time on transesterification of beef tallow with methanol. The reaction was very slow during the first minute due to mixing and dispersion of methanol into beef tallow. From 1 to 5 min, the reaction proceeds very fast. The production of BTME reached the maximum value at about 15 min. Transesterification can occur at different temperatures, depending on the oil used. Recently, Ferdous et al. (2013) studied optimization of biodiesel production from mixed feedstock oil under the condition of methanol–oil molar ratio 6:1, 0.5 percent sodium methoxide catalyst, and 60°C. An approximate yield of 98 percent was observed after 75 min for mixed feedstock oil (Figure 3.9a).

Temperature has a significant effect on conversion of FFA to methyl ester. For example, higher reaction temperature increases the reaction rate and shortens the reaction time due to the reduction in viscosity of oils. Recently, Ferdous et al. (2013) studied the effect of temperature on the

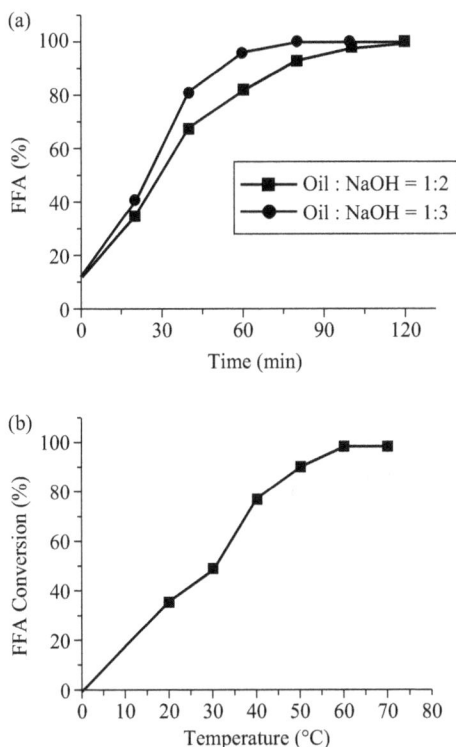

Figure 3.9 Effect of (a) reaction time and (b) temperature on FFA conversion.

conversion mixed feedstock oil to biodiesel. The conversion was of 98 percent and was obtained at 60°C temperature (Figure 3.9b). Further, increase of temperature does not increase the conversion of FFA. Leung and Guo (2006) and Eevera, Rajendran, and Saradha (2009) found that increase in reaction temperature beyond the optimal level leads to decreased biodiesel yield, because higher reaction temperature accelerates the saponification of triglycerides. Usually the transesterification reaction temperature should be below the boiling point of alcohol in order to prevent alcohol evaporation. A maximum yield of ester was obtained in the temperatures ranging from 50°C to 60°C at an alcohol to oil molar ratio of 6:1 (Freedman, Pryde, and Mounts 1984; Leung and Guo 2006; Ma and Hanna 1999).

3.3.5 MIXING INTENSITY

To achieve perfect contact between the reagent and oil during transesterification, they were mixed together. It has been observed that during the transesterification reaction, the reactants initially form a two-phase liquid system. The mixing effect has been found to play a significant role in the slow rate of reaction. As phase separation ceases, mixing on the kinetics of the transesterification process forms the basis for process scale up and design. Influence of mass transfer on the production of biodiesel may be observed through mixing variation as the use of different mixing methods (magnetic stirrer, ultrasound, and ultra turrax) results in different conversions for the transesterification of rape oil with methanol in both acidic and basic systems. The effect of stirring on FAME production was investigated by Rashid and Anwar (2008) in three experiments by differing stirring rates (180, 360, and 600 rpm). In all the experiments, an oil/methanol molar ratio of 1:6, a reaction temperature of 60°C, and a NaOCH$_3$ catalyst concentration of 1.00 percent were used. It was concluded that the mixing rate of 600 rpm afforded the optimum conversion of safflower oil to FAME (98 percent). When agitation speeds are in the range of 400 to 600 rpm and mass transfer limitations are practically eliminated, temperature becomes the most influential factor affecting the apparent rate of transesterification as the system becomes kinetically controlled (Vicente, Martınez, and Aracil 2004; Poljanšek and Likozar 2011). It was reported by Zhang, Stanciulescu, and Ikura (2009) that the conventional base-catalyzed transesterification is characterized by slow reaction rates at both initial and final reaction stages limited by mass transfer between polar methanol/glycerol phase and non-polar oil phase.

3.3.6 EFFECT OF USING ORGANIC CO-SOLVENTS

The reaction of fatty acids with methanol is reversible; the reaction comes to equilibrium before a complete conversion of the oil. This is why an extent amount of methanol is added to the reaction mixture, in order to shift the equilibrium to the product side. Adding more methanol is not a solution by itself, since for complete reaction impractical amounts of methanol may be needed for a high conversion degree. The rate of reaction decreases as it approaches equilibrium. The main problem of the transesterification reaction is that the reactants are not readily miscible. This lowers the rate of collisions of molecules and so the rate of reaction causes longer reaction times, higher operating expenses, and labor. To overcome this difficulty of the heterogeneous mixing of the reactants, a single phase reaction has been proposed by Boocock et al. (1998). The proposed model includes a cyclic solvent introduced into the reaction mixture, which makes both the oil and methanol miscible. This solvent can have numerous different solvents with the boiling point up to 100°C. Tetrahydrofuran (THF) is preferred because of its close boiling point to that of methanol so that after reaction both methanol and THF can be recycled in a single step to use again.

In order to conduct the reaction in a single phase, co-solvents like THF were tested by Boocock et al. (1996). At the 6:1 methanol–oil molar ratio, the addition of 1.25 volume of THF per volume of methanol produces an oil dominant one phase system in which methanolysis speeds up dramatically and occurs as fast as butanolysis. In particular, THF is chosen because of its boiling point of 67°C is only two degrees higher than that of methanol. Therefore, at the end of the reaction the unreacted methanol and THF can be co-distilled and recycled. Krisnangkura and Simamaharnnop (1992) studied the continuous transmethylation of palm oil using toluene as a co-solvent. The highest conversion of 96 percent was obtained within 60 sec at the methanol to oil molar ratio of 13:1. It was reported that benzene was a good solvent for transmethylation, but the yield of palm oil methyl ester was slightly lower than toluene (Krisnangkura and Simamaharnnop 1992).

Pardal et al. (2010) studied the transesterification of rapeseed oil using various co-solvents including diethyl ether, dibutyl ether, tert-butyl methyl ether, diisopropyl ether, THF, and acetone. It was reported that generally a molar ratio 1:1 between methanol and the co-solvent is enough for assuring a good conversion. The highest yield of 97.6 percent was obtained when using 0.7 percent of KOH as catalyst, a molar ratio of methanol to oil 9:1, a molar ratio of methanol to diethyl ether 1:1, a reaction temperature of 303 K, 700 rpm, and 120 min of reaction. Thus, the co-solvent enhances the transesterification reaction, as reported by Lam et al. (2010). The addition of biodiesel as a co-solvent could reduce the methanol needed for completing the reaction. Moreover, it could reduce the amount of catalyst needed for the reaction to take place, and the negative effects associated with caustic such as production of soaps. Thus, the utilization of biodiesel as a co-solvent, which is part of the process on a large scale, simplifies the operations of the biodiesel plant.

3.4   HETEROGENEOUSLY CATALYZED PROCESS

Heterogeneously catalyzed transesterification reaction is complex because it occurs in a three-phase system consisting of a solid (heterogeneous catalyst) and two immiscible liquid phases (oil and methanol). The need for development of heterogeneous catalysts has arisen from the fact that homogeneous catalysts used for biodiesel development pose a few draw-backs discussed in previous sections. Easy separation, easy recovery, no problems in solubility, and miscibility are the strengths of a heterogeneous system in order to reduce the cost of production. Heterogeneous catalysis is thus considered to be a green process. Needless to say, because of these advantages, research on the transesterification reaction using heterogeneous catalysts for biodiesel production has increased over the past decade. The size of the heterogeneous catalyst employed for biodiesel production was from nano to macrometer in ranges (Islam et al. 2013b), as shown in Figure 3.10.

A great variety of materials have been tested as heterogeneous catalysts for the transesterification of vegetable oils, as shown in Table 3.1. It was reported by several researchers (Di Serio et al. 2007a, 2007b; Islam et al. 2013a; Lotero et al. 2005) that the efficiency of the catalyst depends on several factors such as specific surface area, pore size, pore volume, acidity or basicity, and active site concentration of catalyst (Figure 3.11). Besides, the numbers of operating parameters such as temperature, extent of catalyst loading, mode of mixing, alcohol/oil molar ratio, presence/absence of impurities in the feedstock, and the time of reaction are important.

Several industrial biodiesel manufacturers have already adopted heterogeneous processes proving that a considerable amount of progress was made in this direction. For instance, Axens–IFP Group Technologies commercialized Esterfip-H, an innovative heterogeneously catalyzed technology for the production of high-quality biodiesel that also allows the production of good-quality glycerol. Two Esterfip-H plants in France and Sweden have been built so far, and several other plants are in preparation (Lengyel, Cvengrošová, and Cvengroš 2009; Van Walwijk 2005). NOVA Biosource Fuels offers a patented, heterogeneous catalytic conver-

Figure 3.10. Catalyst employed for biodiesel production (a) macro catalyst and (b) nano catalyst.

Table 3.1. Heterogeneous catalysts employed for transesterification reactions

Catalyst type		Example
Solid acid catalyst	Zeolite type solid acid catalyst	Zeolite socony mobil-5 (HZSM-5); zeolite-β; zeolites-Y
	Heteropoly acid loaded MCM-41 catalyst	Mg-mobile crystalline materi-al-41(MCM-41); Al-MCM-41
	Sulfated zirconia and tin oxide type solid acid catalyst	SO_4^{2-}/ZrO_2; SO_4^{2-}/SnO_2
	Tungsten trioxide loaded zirconia type solid acid catalyst	WO_3/ZrO_2
Solid basic catalyst	Alkali metal salts loaded on alumina	KI/Al_2O_3; $Mg(NO_3)_2/Al_2O_3$ $Na/\gamma\text{-}Al_2O_3$; $Na/NaOH/\gamma\text{-}Al_2O_3$; $NaOH/\gamma\text{-}Al_2O_3$; KF/Al_2O_3; KCO_3/Al_2O_3; KNO_3/Al_2O_3; $LiNO_3/Al_2O_3$; $Ca(NO_3)_2/Al_2O_3$; $NaNO_3/Al_2O_3$; KOH/Al_2O_3
	Alkaline earth metal oxide	MgO; CaO; $CaCO_3$; $Ca(OH)_2$ SrO; $CaO\text{--}La_2O_3$,
	Hydrotalcites	MgO/Al_2O_3; Mgo/CaO
	Zeolites	ETS-10; KOH/NaX; NaO/NaX; CsX; KI/NaX

sion process to treat both FFAs and triglycerides (Salzano, Di Serio, and Santacesaria 2010).[1] It may not be possible with any one of the catalysts to possess simultaneously a strong acid/base, high surface area, porous, and inexpensive catalyst production. Certainly, one needs to strike a balance by considering all the process parameters in each case. The extensive effort on the development of heterogeneous catalysts for biodiesel production, as summarized, has led to an enhanced understanding on the chemical and

---

[1]Goswami, A., & Block, D. S. An overview of biodiesel transesterification tech-nology in India. In *The outcome of the ESCAP-APCAEM forum serves to guide the formulation of capacity-building programmes for policymakers, development practitioners and CDM projects on bioenergy. The forum also provided a platform for exchange of best practices and innovative solutions on how to foster public/private partnerships that would promote bioenergy* (p. 27).

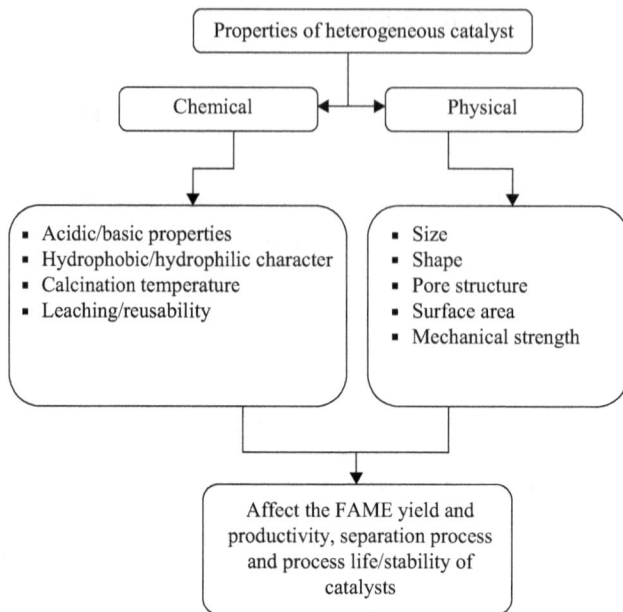

Figure 3.11. Design consideration for the catalyst properties for the transesterification reaction.

physical properties of a catalyst that play a vital role on the biodiesel yield. Basically, base catalysis is a better choice than acid catalysis in terms of the reaction rate and biodiesel productivity. However, the adverse effect on the activity by water and FFA must be overcome before any developed catalyst can claim to possess robust activity for base-catalyzed biodiesel synthesis. Though, the high yield of biodiesel has been produced using solid base or acid catalyst, the catalysts are in the form of powders with diameter ranging from nano- to micrometer. From the practical point of view, handling of small particles could be difficult due to the formation of pulverulent materials. Utilization of powders in catalytic reactions renders their recovery and purification, and energy-intensive ultracentrifugation is needed for the subsequent separation operation. In addition, the active phase of the powdered catalyst may not be uniformly distributed on the support but rather forms localized aggregates leading to low contact of active surface in the catalyst. Thus, the efficiency of the catalyst and its feasibility at industrial scale might be reduced. Based on the above discussions, developing a new solid the design of a catalyst at a millimetric seems to be an appropriate solution to overcome problems associated with the traditional catalysts. The mechanical strength and shape of the catalyst is the key issue for the millimetric heterogeneous catalysts. As previously stated, supported catalyst in spherical form can offer shape-dependent advantages such as minimizing the abrasion of catalyst in the reaction environment. It is further highlighted that high mechanical strength is crucial for the long-term stability of the catalyst. Therefore, easy handling, separation, and reusability are the main strengths that could response to select the macro catalyst.

3.4.1 HETEROGENEOUS ACID-CATALYZED PROCESS

Solid acid catalysts have the potential to replace strong liquid acids to eliminate the corrosion problems and consequent environmental hazards posed by the liquid acids. However, the efforts at exploiting solid acid catalysts for transesterification are limited due to the pessimistic expectations on the possibility of low reaction rates and adverse side reaction. Since heterogeneous catalysis is a surface phenomenon, the extend to acidity of a catalyst affects its catalytic activity, simply because of the variation in the number of acid sites. As a result of the development of solid acid catalyst for biodiesel production, the relationship between catalytic activity and strength of acidity has been the subject of an increasing number of investigations in recent years. Super acidic sulfated zirconia (SO_4/ZrO_2), tungstated zirconia (WO_3/ZrO_2), sulfated tin oxide (SO_4^{2-}/SnO_2), and tungstated zirconia-alumina (WZA) (Furuta, Matsuhashi, and Arata 2004; Furuta, Matsuhashi, and Arata 2006; Matsuhashi et al. 2001; Yadav and Murkute 2004) have been evaluated for the transesterification of vegetable oil at a temperature in the range of 150°C to 250°C. Among them, WO_3/ZrO_2 catalyst was found to be most favorable to the transesterification reaction and resulted in more than 90 percent yield at 200°C with 15:1 methanol to oil ratio. The catalyst was found to be suitable for transesterification reaction due to the coexistence of tetragonal zirconia with the amorphous tungsten oxide. Recently, the activity of several acid catalysts using a different ratio of TiO_2/SO_4 synthesized via sol-gel technique in the transesterification of soybean oil with methanol at 120°C has been ranked in reference to their ratio of TiO_2/SO_4 such as $TiO_2/SO_4(5{:}1)$ > $TiO_2/SO_4(10{:}1)$ > TiO_2/SO_4 (20:1) (De Almeida et al. 2008). Solid acids keep stable activity in conversion of low-qualified oils or fats to biodiesel. Currently, developed solid acid catalysts are introduced in the following sections: cation exchange resin (i.e., Amberlyst-15 and NR50), mineral salts (i.e., ferric sulfate, zirconium sulfate, alum phosphate, and zirconium tungsten), supported solid acid, and heteropoly acid catalysts. A heterogeneous acid-catalyzed reaction is slower than the heterogeneous base-catalyzed reaction. Therefore, many researchers have been directed toward the development of a heterogeneous basic catalytic system aiming to reduce the reaction temperature.

3.4.2 HETEROGENEOUS BASE-CATALYZED PROCESS

Solid-base catalysts have been applied in transesterification reaction of a variety of supported materials, such as alkaline-earth metal oxides and hydroxides, alkali metals hydroxides, or salts supported on alumina, zeolites, and hydrotalcites. The KNO_3, K_2CO_3, KOH, KI supported on Al_2O_3 catalysts showed high yield of biodiesel (>90 percent) due to more basic sites forming either K_2O supported on Al_2O_3 species produced by thermal decomposition or Al–O–K groups formed by salt–support interactions (Xie and Li 2006; Xie, Peng, and Chen 2006). Ma et al. (2008) and Kim et al. (2004) investigated the activity of $K/KOH/\gamma$-Al_2O_3 and $Na/NaOH/\gamma$-Al_2O_3 catalyst in the transesterification reactions. It was pointed out that the catalyst's activity is closely related to the basic nature of the catalyst. The strong basic sites (super basic) promote the transesterification reaction at low temperature (60°C to 70°C), while the basic sites with medium strength require a higher temperature to process the reaction. Since the ability of bases to abstract a proton from an alcohol is directly connected to the base strength, stronger bases are in general more effective to initiate the transesterification of triglycerides. It was reported by Chorkendorff and Niemantsverdriet (2006) that metal oxide provides sufficient adsorptive sites for alcohol in transesterification reaction and concluded that the high transesterification activity of catalyst might be due to the manifestation of the dissociation of alcohol to RO^- and H^+ on basic sites of metal oxide catalyst surface. Thus, the generation of active oxide phases such as K_2O phases on $K/KOH/\gamma$-Al_2O_3 and $NaAlO_2$ phase on $NaNO_3/\gamma$-Al_2O_3 increased the basicity as well as the transesterification activity, as reported by Ma et al. (2008). The catalyst was found most promising for higher FAME yield of 95 percent; however, a small portion of metal content such as K^+ or Na^+ leaching was observed in methanol. A number of process variables affect the efficiency of transesterification reaction, which will be described in the subsequent section. Heterogeneous base catalysis has a shorter history than that of heterogeneous acid catalysis. Solid bases refer mainly to solids with Brønsted basic and Lewis basic activity centers, which can supply electrons (or accept protons) for (or from) reactants. Heterogeneous base-catalyzed transesterification for biodiesel synthesis has been studied intensively over the last decade. Low-qualified oil or fat with FFAs and water can be used. However, the catalytic efficiency of conventional heterogeneous base catalysts is relative low and needs to be improved. Various types of catalytic materials have been studied to improve the transesterification of glycerides.

3.5 PROCESS VARIABLES OF A HETEROGENEOUSLY-CATALYZED SYSTEM

There are a number of factors or variables that could affect the transesterification process. The factors or variables usually have different effect on the transesterification process depending on the catalyst used for the transesterification process. Optimization of these process variables is usually carried out by changing the certain variables while maintaining the other variables at fixed values and then subsequently comparing the yield or conversion of FAMEs. Process variables, such as molar ratio of alcohol to oil, catalyst amount, reaction time, mixing intensity, reaction temperature, catalyst particle size, and shape of particles are introduced here in detail.

3.5.1 EFFECT OF MOLAR RATIO OF ALCOHOL TO OIL

One of the most important factors that affect the yield of ester (biodiesel) is the molar ratio of alcohol to triglyceride (oil). Based on the stoichiometric of transesterification reaction, every mol of triglyceride requires three moles of alcohol to produce three moles of fatty acid alkyl esters and one mole of glycerol. Transesterification is an equilibrium reaction in which an excess of alcohol is required to drive the reaction to the right (Ma and Hanna 1999). However, an excessive amount of alcohol makes the recovery of the glycerol difficult, so that the ideal alcohol/oil ratio has to be established empirically (Schuchardt, Sercheli, and Vargas 1998).

Transesterification of rapeseed oil carried out by Kawashima, Matsubara, and Honda (2008) showed a maximum conversion at 6:1 of methanol to oil ratio, whereas an earlier study by Xie, Peng, and Chen (2006) found a maximum conversion at a ratio of 15:1. With further increase in molar ratio, the conversion efficiency more or less remains the same. Most of the studies on the solid base-catalyzed transesterification of vegetable reported that maximum conversion to the ester occurred with a molar ratio of 6:1 (Demirbas 2007a; Kawashima, Matsubara, and Honda 2008; Kim et al. 2004; Portnoff et al. 2005). However, some other results showed that the optimum molar ratio of oil to alcohol was 9:1 (Albuquerque et al. 2008), 40:1 (Tateno and Sasaki 2004), 12:1 (Albuquerque et al. 2008), 15:1 (Xie and Li 2006), and 30:1 (Ngamcharussrivichai, Totarat, and Bunyakiat 2008) to get the maximum biodiesel yield. Solid acid-catalyzed reactions require the use of high alcohol-to-oil molar ratios in order to obtain good product yields in practical reaction times. Higher molar ratios showed only moderate improvement until reaching a maximum value at a 55:1 ratio (82 percent) (Antunes, Veloso, and Henriques 2008). Therefore, a higher molar ratio of oil to alcohol (>6:1) could also be used as the optimum ratio for oil to methanol, depending on the quality of feedstock and catalyst type of the transesterification process.

3.5.2 EFFECT OF CATALYST AMOUNT

The type and amount of catalyst required in the transesterification process usually depends on the quality of the feedstock applied for the transesterification process. For a purified feedstock, any type of catalyst could be used for the transesterification process. However, for feedstock with high moisture and FFAs content, the homogenous transesterification process is unsuitable due to high possibility of saponification process instead of transesterification process (Gerpen 2005). The yield of fatty acid alkyl esters generally increases with increased amount of catalyst (Demirbas 2007a; Ma and Hanna 1999). This is due to the availability of more active sites by additions of a larger amount of catalyst in the transesterification process. However, the addition of an excessive amount of catalyst gives rise to the formation of an emulsion, which increases the viscosity and leads to the formation of gels (Encinar, Gonzalez, and Rodríguez-Reinares 2005). These hinder the glycerol separation and, hence, reduce the apparent ester yield. Therefore, similar to the ratio of oil to alcohol, the optimization process is necessary to determine the optimum amount of catalyst required in the transesterification process.

In the case of the heterogeneous catalysis, the literature presents many works relating to this issue. In most of the literature reviewed, the results showed that the best suited catalyst concentrations giving the best yields of the esters are between 2.5 and 10 wt% (Benjapornkulaphong, Ngamcharussrivichai, and Bunyakiat 2009; Boz, Degirmenbasi, and Kalyon 2009; Noiroj et al. 2009; Samart, Sreetongkittikul, and Sookman 2009). Xu et al. (2009) used Ta_2O_5/SiO_2-$[H_3PW_{12}O_40/R]$ (R = Me or Ph) as the catalyst in the transesterification of soybean oil and reported that 2 wt% (in terms of oil) catalyst is the optimum catalyst concentration. Similarly, Xie, Yang, and Chun (2007) carried out transesterification of waste rapeseed oil and obtained maximum conversion at 3 wt% NaX/KOH catalyst concentration and further increase in the catalyst concentration had no effect on conversion.

These results were confirmed by Xie, Peng, and Chen (2006) who carried out transesterification of soybean oil with KI/Al_2O_3 concentrations at 0.05 wt% increments starting from 1 to 3.5 wt% and observed that the highest conversion was achieved at 2.5 wt% concentration.

3.5.3   EFFECT OF REACTION TIME

For a heterogeneous transesterification process, the reaction period varies depending on the reactivity and type of the solid catalyst used. For a practical and economic feasible transesterification process, it is necessary to limit the reaction time at a certain period. A longer reaction time could also permit reversible transesterification reaction to occur, which eventually could reduce the yield of fatty acid alkyl esters (Demirbas 2009a, 2009b). Thus, optimization of reaction time is also necessary.

Most investigators have observed an optimum reaction time for a basic-catalyzed transesterification process around 3 to 12 hrs (Benjapornkulaphong, Ngamcharussrivichai, and Bunyakiat 2009; Boz, Degirmenbasi, and Kalyon 2009; Noiroj et al. 2009). However, a direct comparison of the biodiesel yield with the reaction time is difficult because of the variation in other reaction conditions, notably, the oil to methanol molar ratio and amount of catalyst used. Current researches have shown that the reaction time for a non-catalytic transesterification process using supercritical alcohol is shorter compared to conventional catalytic transesterification process (Demirbas 2007). However, the non-catalytic transesterification process using supercritical alcohol is much more energy intensive than the solid base-catalyzed process, because it operates at very high pressures (200 to 450 bar) and the high temperatures (350 to 400°C) bring along proportionally high heating and cooling costs. It was also reported that excess reaction time does not increase the conversion but favors the backward reaction (hydrolysis of esters), which results in a reduction of product yield (Leung and Guo 2006).

3.5.4   EFFECT OF REACTION TEMPERATURE

Temperature is an important parameter as it allows the faster reaction kinetics and mass transfer rates in the transesterification reaction (Liu 1994). Normally, a relatively high reaction temperature is required for heterogeneous system in order to increase the mass transfer rate between reactant molecules and the catalyst. This is due to the existence of the initial 3-phase mixture: oil–methanol–solid catalyst. Higher temperatures decrease the time required to reach maximum conversion (Pinto et al. 2005).

Many studies have shown that reaction temperature significantly influences FAME yield for transesterification reaction catalyzed by heterogeneous process. Boz, Degirmenbasi, and Kalyon (2009) found that the yield of biodiesel was tripled (30 to 99 percent) using solid base catalyst (KF/γ-Al$_2$O$_3$) when the transesterification temperature increased from 25°C to 65°C. Similar results were reported by Samart, Sreetongkittikul, and Sookman (2009). The conversion was increased from 68 percent at 50°C to 90 percent at 70°C with a 16:1 molar ratio of methanol to oil using KI/mesoporous silica catalyst. The influence of FAME yield on TiO$_2$/SiO$_2$ as a solid acid-catalyzed transesterification has been studied by Di Serio et al. (2007). Their study shows that the yield of FAME was increased from 5 to 62 percent when the temperature was changed from 120°C to 180°C for transesterification of soybean oil with a molar ratio of methanol to oil of 1/1 (w/w). Similarly, Lam et al. (2010) showed that the transesterification yield of waste cooking oil increased from 80.3 to 88.2 percent as the temperature increased from 60°C to 100°C. The methanol to oil molar ratio was 15, the catalyst (SO$^{2-}_4$/SnO$_2$-SiO$_2$) (referring to weight of oil) was 6 wt%, and 1 hr reaction time was used.

Researchers have found that the reactions are accelerated at critical point conditions. The critical temperatures and critical pressures of the various alcohols are shown in Table 2.1. Madras, Kolluru, and Kumar (2004) showed that transesterification conversion of sunflower oil increased from 78 to 96 percent as the temperature increased from 200°C to 400°C. The methanol to oil molar ratio was 40, the pressure was 200 bar, and a 40 min reaction time was used. Similar results were reported by Demirbas (2002). The conversion at 5 min can be nearly doubled from 50 percent at 177°C to over 95 percent at 250°C with a 41:1 molar ratio of methanol to hazelnut kernel oil. Other results consistent with this finding are from Demirbas (2003). Their study shows that that the yield of FAME increased (from 5 to 99 percent) when the transesterification temperature increased from 127°C to 337°C as shown in Figure 3.12. Thus, the temperature had a favorable effect on FAME yield.

Transesterification can be conducted using solid catalyst at various temperatures ranging from 60°C to 450°C. However, the operating temperature for transesterification process depends on the method used. Certain processes, heterogeneous acid catalyzed reaction transesterification process, generally require moderate temperature ranging from 120°C to 250°C (De Almeida et al. 2008). However, the non-catalytic transesterification process requires high temperature ranging from 230°C to 450°C to yield the desired product (fatty acid alkyl esters) (Demirbas 2007a). A great variety of solid basic catalysts such as alkaline-earth metal oxides

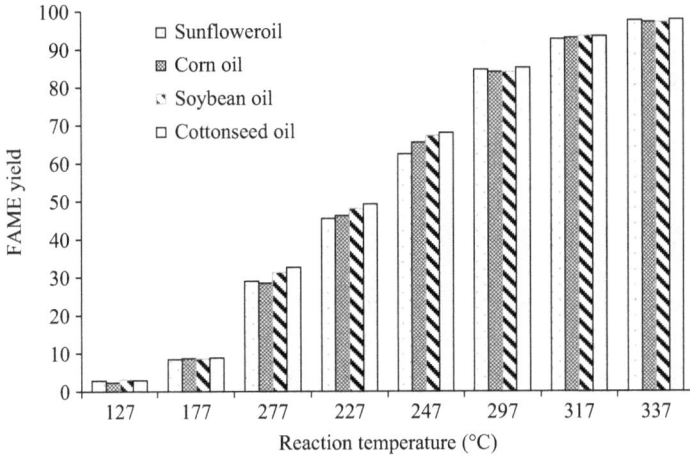

Figure 3.12. Changes in yield percentage of methyl esters as treated with supercritical methanol at different temperatures as a function of reaction temperature (Demirbas 2003).

and hydroxides, alkali metal hydroxides or salts supported on alumina, zeolites, and hydrotalcites have been evaluated to date, which have shown to be good candidates for transesterification reaction at relatively low temperature ranges (60°C to 70°C) (MacLeod et al. 2008; Bo et al. 2007).

3.5.5 EFFECT OF MIXING INTENSITY

Mixing is very important in the transesterification process, as oils or fats are immiscible with alcohol. As a result, vigorous mixing is required to increase the area of contact between the two immiscible phases (Singh, Fernando and Hernandez 2007; Meher, Vidya Sagar, and Naik 2006). Mechanical mixing is commonly used in the transesterification process. The intensity of the mixing could be varied depending on its necessity in the transesterification process. In general, the mixing intensity must be increased to ensure good and uniform mixing of the feedstock. When vegetable oils with high kinematic viscosity are used as feedstock, intensive mechanical mixing is required to overcome the negative effect of viscosity to the mass transfer between oil, alcohol, and catalyst.

Poor mass transfer between two phases in the initial phase of the reaction results in a slow reaction rate, the reaction being mass transfer controlled (Noureddini and Zhu 1997). Stamenković et al. (2007) studied the effect of agitation intensity on alkali-catalyzed methanolysis of sunflower oil and reported that the drop size distributions of emulsion were found to become narrower and shift to smaller sizes with increasing agitation speed. It is evident, from the literature presented earlier, that the agitation had a favorable effect on FAME yield. Therefore, variations in mixing intensity are expected to alter the kinetics of the transesterification reaction.

3.5.6 EFFECT OF CATALYST PARTICLE SIZE

The catalytic activity of the materials has been reported to be dependent on the particle size. Smaller particles can be expected to exhibit a higher rate of reaction, or consequently conversions for a given volume of reaction mass due to increased external surface available (McCarty and Weiss 1999).

Gutierrez-Ortiz et al. (2000) studied in detail the catalytic hydrogenation of methyl oleate, which is the main component of olive oil, using a Ni/SiO$_2$ catalyst in a slurry reactor. The authors concluded that, at six bars and 180°C, both activity and selectivity significantly decreased when the size of the catalyst particles were larger than 50 μm and the stirring rate was below 2000 rpm. Şensöz, Angın, and Yorgun (2000) investigated the influence of particle size on the pyrolysis of rapeseed by varying the particle size of rapeseed in the range of 0.224 to 1.8 mm and found that the yields of products are largely independent of particle size. Other results consistent with these findings are from Ferretti et al. (2009). In order to investigate the effect of the MgO particle size on FAME conversion, they carried out several 3-hr catalytic tests using three different particle size ranges (100, 100 to 177, and 177 to 250 μm), without changing any other reaction parameter. Only small differences were observed during the 3-hr tests. The two catalytic tests with the smallest particles show slightly lower FAME conversions than the experiment with the largest size. This result is the opposite of that expected in the presence of diffusional limitations. The authors concluded that this effect could be attributed to a "flotation effect" of the smallest particles in the presence of the foam caused by the surfactant monoglyceride that probably places the catalyst surface far from the glycerol phase, thereby decreasing the FAME conversion. Therefore, for practical reasons and to avoid flotation of small particles, the largest particle size range has been adopted for the glycerolysis of fatty acid ethyl esters using MgO catalyst.

Recent advances in nanoscience and nanotechnology have led to a new research interest in employing nanometer-sized particles as an alternative matrix for supporting catalytic reactions. Compared with conventional supports like solid-phase, nanoparticular matrices could have a higher catalyst loading capacity due to their very large surface areas (McCarty and Weiss 1999). Freese, Heinrich, and Roessner (1999) have reported the catalytic activity of micrometer zeolite sieve of molecular porosity (ZSM-5) catalysts and found that the catalyst exhibited a conversion of 14.2 percent, which is much lower than the nanocrystalline ZSM-5 catalysts used in their investigation. The higher activity of nanocrystalline ZSM-5 is probably due to the increased external surface of smaller crystal.

Mabaso, Van Steen, and Claeys (2006) reported the effect of crystal size on carbon supported iron catalysts prepared via precipitation where catalysts with smaller metal than 7 to 9 nm have showed higher selectivity of methane conversion compared to the bigger-sized catalysts. However, the effect of particle size on FAME has been studied to a lesser extent. Recently, the utilization of Al_2O_3 supported KF catalyst having a particle size in the order of nanometer order for biodiesel production has been demonstrated by Boz, Degirmenbasi, and Kalyon (2009) and reported that the catalyst can be used in the production of biodiesel from vegetable oil. The high yield of FAME (>90 percent) has been achieved using the catalyst ranging from nanometer to micrometer in diameter. However, from the practical point of view, handling of such small particles in large quantities could be difficult due to the formation of dust.

The smaller the emitted particle, the more harmful it is to the human body because particles under 100 nm (ultrafine particles) in diameter have a higher surface area per unit mass of particles; therefore, the smaller particles can more easily infiltrate into the respiratory organs (Donaldson, Li, and MacNee 1998). Utilization of powders in conventional catalytic reactions is problematic because powder form catalysts are at a disadvantage in pressure drop, mass/heat transfer, contacting efficiency, and separation processes (Centi and Perathoner 2003). Therefore, the design of a catalyst's form at a macroscale is indispensable to avoid these problems. From these viewpoints, macrostructured materials have drawn attention as catalytic supporting materials (Centi and Perathoner 2003). It was reported by Jarrah, van Ommen, and Lefferts (2004) that the macroscopic particle will open-up a real opportunity for their use as a catalyst support in relation to the traditional catalyst carriers. Among the different potential applications of these materials, catalysis either within the gas or the liquid phase seems to be the most promising according to the results recently reported in literature.

3.5.7 SHAPE OF PARTICLES

The shape of the particles was measured quantitatively by means of sphericity factor (SF) as described by Chan et al. (2009). The SF provides brief classification about the degree of deviation of the irregular particle from the true sphere shape with zero as perfect sphere and the increasing value indicates a higher degree of deformation. It has been shown by Chan et al. (2009) that a particle with SF less than 0.05 can be considered as spherical. In addition, supported catalyst in spherical form can offer shape-dependent advantages such as minimizing the abrasion of catalyst in the reaction environment as reported by Campanati, Fornasari, and Vaccari (2003). In fact, the stability of the catalyst will be increased.

3.6 ENZYME-CATALYZED TRANSESTERIFICATION

The development of green processes for biodiesel production has received much attention, involving the use of heterogenous catalysts, either chemical or enzymatic. In this sense, enzymatic processes can provide significant advantages over chemical process for biodiesel production. The mild conditions (pH, temperature, and pressure), usually applied in the biocatalyzed reaction, allow energy saving and product quality improvement, because of the minimal thermal degradation of the substrates (Dossat, Combes, and Marty 1999; Fjerbaek, Christensen, and Norddahl 2009; Kumari et al. 2009). The enzymes can be used in solution or immobilized onto a support material, which allows the use of fixed-bed reactors. The reaction can be performed at 35°C to 45°C. However, the reaction is very slow, requiring from 4 to 40 hrs. Because of the high cost of the enzymes, this process is not economically feasible for biodiesel production at this time. Abdulla and Ravindra (2013) reported a supported enzyme catalyst used for biodiesel production as shown in Figure 3.13.

Due to the applicability of enzymes for biodiesel production regardless of large variations in the quality of the raw material, enzymes can have an industrial potential, which is worth further elaboration, because

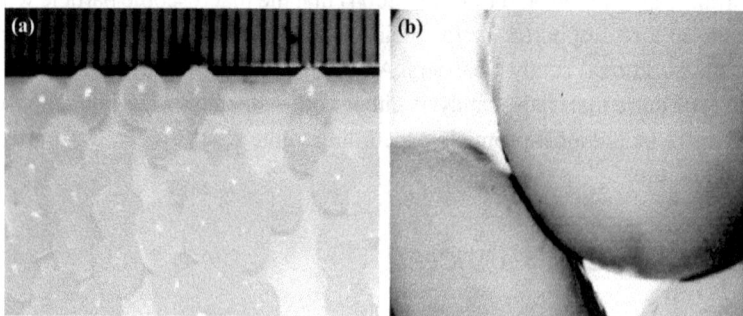

Figure 3.13. The supported enzyme catalyst supported on (a) Na-alginate and (b) K-carrageenan used for biodiesel production.

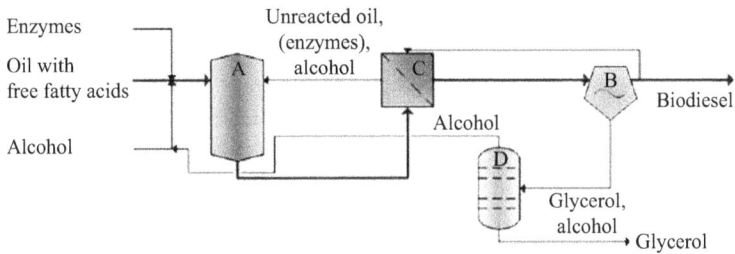

Figure 3.14. Ideal process design for enzymatic biodiesel production.

of the advantages named in the introduction compared to the traditional two-step process with chemical catalysts. An ideal process design for enzymes catalyst biodiesel production reported by Fjerbaek, Christensen, and Norddahl (2009) as shown in Figure 3.14.

There is a current interest in using enzymatic catalysis to commercially convert vegetable oils and fats to FAME as biodiesel fuel, since it is more efficient, highly selective, involves less energy consumption (reactions can be carried out in mild conditions), and produces less side products or waste (environmentally favorable) (Akoh et al. 2007). The transesterification process is catalyzed by lipases such as *Candida antarctica* (Royon et al. 2007), *Candida rugosa* (Linko et al. 1998), *Pseudomonas cepacia* (Ghanem 2003), immobilized lipase (Bernardes et al. 2007), and *Pseudomonas* sp. (Ming, Ghazali, and Chiew Let 1999).

The enzymatic alcoholysis of soybean oil with methanol and ethanol was investigated using a commercial, immobilized lipase (Bernardes et al. 2007). In that study, the best conditions were obtained in a solvent-free system with ethanol/oil molar ratio of 3.0, temperature of 50°C, and enzyme concentration of 7.0% (w/w). They obtained yield 60 percent after 1 hr of reaction. In another study, Shah and Gupta (2007) obtained a high yield (98 percent) by using *P. cepacia* lipase immobilized on celite at 50°C in the presence of 4 to 5% (w/w) water in 8 hrs. A more recent study by Maceiras et al. (2009) was also conducted to investigate the enzymatic conversion of waste cooking oils into biodiesel using immobilized lipase Novozym 435 as catalyst. The effects of methanol to oil molar ratio, dosage of enzyme and reaction time were investigated. The optimum reaction conditions for fresh enzyme were methanol to oil molar ratio of 25:1, 10 percent of Novozym 435 based on oil weight, and reaction period of 4 hrs at 50°C obtaining a biodiesel yield of 89.1 percent. Moreover, the reusability of the lipase over repeated cycles was also investigated under standard conditions.

Tamalampudi et al. (2008) employed immobilized whole cell and commercial lipase as biocatalyst for biodiesel production from Jatropha oil. The lipase producing whole cells of *Rhizopus oryzae* (ROL) immobilized onto biomass support particles (BSPs) were used for the production of biodiesel from relatively low-cost nonedible oil from the seeds of *J. curcas*. The activity of ROL was compared with that of the commercially available, most effective lipase (Novozym 435).

The various alcohols were tested as a possible hydroxyl donor, and methanolysis of jatropha oil progresses faster than other alcoholysis regardless of the lipases used. The maximum methyl esters content in the reaction mixture reaches 80 wt% after 60 hrs using ROL, whereas it is 76 percent after 90 hrs using Novozym 435. Zheng et al. (2009) reported the lipase catalyzed transesterification process in a solvent-free system. Feruloylated diacylglycerol (FDAG) was synthesized using a selective lipase-catalyzed transesterification between ethyl ferulate and triolein. The highest reaction conversion and selectivity toward FDAG were 73.9 percent and 92.3 percent, respectively, at 338 K, reaction time of 5.3 days, with enzyme loading of 30.4 mg/mL; water activity is 0.08, and the substrate molar ratio is 3.7. The disadvantage of the enzyme-catalyzed process is that it is time consuming compared to acid- or base-catalyzed transesterification. Enzyme catalyst have several advantages over chemical catalysts such as mild reaction conditions, specificity, and reuse; and enzymes or whole cells can be immobilized, are considered natural, and the reactions they catalyze are considered green reactions (Akoh et al. 2007). However, the drawbacks of enzymatic catalysts have significantly higher production costs as well as cause difficulty during manufacturing due to the need for a careful control of reaction parameters. Moreover, the reaction yields as well as the reaction times are still unfavorable compared to the alkaline-catalyzed reaction systems.

Several researchers (Fjerbaek, Christensen, and Norddahl 2009; Akoh et al. 2007; Iso et al. 2001) compared the enzymatic process with the alkaline or acid catalyst.

Enzymes are potentially useful compared to alkaline or acid catalyst, because they are:

- Compatible with variations in the quality of the raw material and reusable
- Able to produce biodiesel in fewer process steps using less energy and with drastically reduced amount of wastewater
- Able to improve product separation and to yield a higher quality of glycerol

Drawbacks for the use of enzymes are:

- Low reaction rate
- Their cost for industrial-scale use 1,000 US$ per kg compared to 0.62 US$ for sodium hydroxide
- Loss of activity, typically within 100 days of operation

These are the key issues to be addressed for the industrial use of lipases in biodiesel production to be viable. Preparation of fatty acid alkyl ester using alternative methods such as ultrasound-assistant process, microwave-assistant process, and non-catalytic process offers a fast, easy route to this valuable biofuel with advantages of a short reaction time, a low reactive ratio, an ease of operation, a drastic reduction in the quantity of by-products, and all with reduced energy consumption. Moreover, this process accelerates the chemical reaction and high yields of biodiesel, as compared to the traditional process that will be described in subsequent sections.

3.7   ULTRASOUND-ASSISTANT PROCESS

In the ultrasonic reactor method, the ultrasonic waves cause the reaction mixture to produce and collapse bubbles constantly. This cavitation provides simultaneously the mixing and heating required to carry out the transesterification process. Thus, using an ultrasonic reactor for biodiesel production drastically reduces the reaction time, reaction temperatures, and energy input. Hence, the process of transesterification can run inline rather than using the time-consuming batch processing. Industrial scale ultrasonic devices allow for the industrial scale processing of several thousand barrels per day. A laboratory scale ultrasonic flow reactor (Gude et al. 2013) is shown in Figure 3.15.

The reasons to consider the ultrasound process is the shortened reaction time and increased biodiesel production. This could be due to the following reasons, as reported by several researchers (Hanh et al. 2009a, 2009b; Mootabadi et al. 2010; Ji et al. 2006; Salamatinia et al. 2010; Teixeira et al. 2009):

- Ultrasound provides the mechanical energy for mixing in which the microturbulence is generated due to radial motion of bubbles leading to intimate mixing of the immiscible reactants, and thus initiating the transesterification reaction.

Figure 3.15. Laboratory scale ultra-sonic reactor.

- The ultrasonic irradiation of a liquid produces acoustic cavitations in which H^+ and OH^- are produced during a transient implosive collapse of bubbles that could accelerate the reaction rate.
- Ultrasonic can also grind the catalyst into smaller particles to create new active sites for the subsequent reaction. Thus, the solid catalyst is expected to last longer in the ultrasonic-assisted process.

However, the higher ultrasound wave amplitude could reduce the yield of biodiesel. It was reported by Choedkiatsakul, Ngaosuwan, and Assabumrungrat (2013) that loss of efficiency in the ultrasonic waves transfers through the liquids may be due to the coalescence of small cavitation bubbles into larger ones which act as a barrier to that wave's transfer and the decoupling effect. Thus, it is very important to optimize ultrasound wave amplitude to get a higher yield of biodiesel.

3.8 MICROWAVE-ASSISTANT PROCESS

Recent laboratory scale microwave applications in biodiesel production proved the potential of the technology to achieve superior results over conventional techniques. Short reaction time, cleaner reaction products, and reduced separation-purification times are the key observations reported by many researchers. Energy utilization and specific energy requirements for microwave-based biodiesel synthesis are reportedly better than conventional techniques. Microwaves can be very well utilized in feedstock preparation, extraction, and transesterification stages of the biodiesel production process. In conventional heating as well as supercritical methods, heat transferred to the sample volume is utilized to increase the temperature of the surface of the vessel followed by the internal materials. This is also called "wall heating." Therefore, a large portion of energy supplied through conventional energy source is lost to the environment through conduction of materials and convection currents. The microwaves provide intense localized heating that may be higher than the recorded temperature of the reaction vessel (Gude et al. 2013) as shown in Figure 3.16.

In recent years, many researchers have tested the application of microwaves in biodiesel production and optimization studies with various feedstock. Transesterification of organic feedstock to yield biodiesel can be performed by the following methods: (1) conventional heating with acid, base catalysts, and co-solvents; (2) sub- and super-critical methanol conditions with co-solvents and without catalyst; (3) enzymatic method using lipases; and (4) microwave irradiation with acid, base, and heterogeneous catalysts. Among these methods, the conventional heating method requires longer reaction times with higher energy inputs and losses to the ambient (Refaat, El Sheltawy, and Sadek 2008). The super and sub-critical methanol process operates in expensive reactors at high temperatures and pressures resulting in higher energy inputs and higher production costs (Demirbas 2002). The enzymatic method, though operates at much lower temperatures, requires much longer reaction times (Fjerbaek, Christensen, and Norddahl 2009; Akoh et al. 2007; Iso et al. 2001). Microwave-assisted

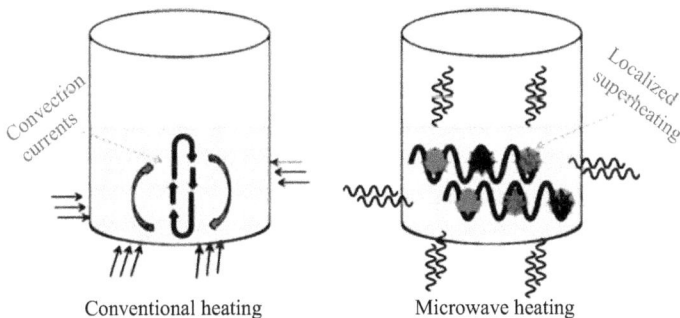

Figure 3.16. Conventional and microwave heating mechanisms.

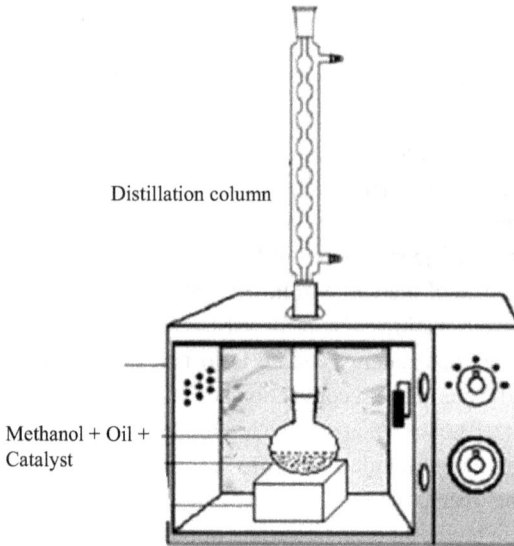

Figure 3.17. Microwave biodiesel production.

transesterification, on the other hand, is energy-efficient and quick process to produce biodiesel from different feedstock. The typical microwave system is shown in Figure 3.17. This process is still in the lab-scale development stage, though the microwave method holds great potential to be an efficient and cost-competitive method for commercial-scale biodiesel production.

The enhanced chemical reaction rate could be due to the following reasons, as speculated by several researchers (Barnard et al. 2007; Di Serio et al. 2007; Leadbeater and Stencel 2006; Lertsathapornsuk et al. 2008; Sheikh et al. 2013)

- Energy transfer from microwaves to the material is believed to occur either through resonance or relaxation, which results in rapid heating and thus, it delivers energy directly to the reactant.
- Microwave assists more molecular friction and collisions in reaction medium, giving rise to intense localized heating and thereby accelerating the chemical reaction.

Understanding the effect of microwaves on biomass extraction and transesterification reactions can be beneficial in the reactor design. Specific areas of challenges that need critical attention prior to large-scale development are controlled heating since biodiesel process is sensitive to temperature variations, efficient transfer of microwave energy into work area with fewer losses to the reactor walls and environment, which includes biodiesel product separation and purification.

3.9 SUPERCRITICAL PROCESS

An alternative, catalyst-free method for transesterification uses supercritical methanol at high temperatures and pressures. In the supercritical state, the oil and methanol are in a single phase, and reaction occurs spontaneously and rapidly. The process can tolerate water in the feedstock, FFAs are converted to methyl esters instead of soap, so a wide variety of feedstock can be used (Minami and Kusdina 2006; Saka and Kusdina 2001). High temperatures and pressures are required, but energy costs of production are similar or less than catalytic production routes (Van Kasteren and Nisworo 2007).

Due to poor methanol and oil miscibility, conversion of oil to biodiesel is a very slow reaction. Use of a co-solvent that is soluble in both methanol and oil may improve reaction rates. The BIOX Process (www.bioxcorp. com) uses either THF or methyl tert-butyl ether (MTBE) as a co-solvent to generate a one-phase system (Ngamprasertsith and Sawangkeaw 2011). In the presence of a co-solvent, the reaction is 95 percent complete in 10 min at ambient temperatures and does not require a catalyst. Deshpande et al. (2010) reported the approach for the production of biodiesel using supercritical method (Figure 3.18). THF has a boiling point very close to that of methanol. The excess methanol and co-solvent are recovered in a single step after the reaction is complete. Co-solvents that are subject to the hazardous and/or toxic air Environmental Protection Agency (EPA) list for air pollutants must be completely removed from the biodiesel and its byproducts (glycerine and methanol). Emissions must be tightly controlled, and processing equipment must be leak proof. The second non-catalytic approach utilizes methanol at very high temperature and pressure (350°C to 400°C and greater than 80 atm or 1200 psi) to convert oil to biodiesel (Demirbas 2002; Cao, Han, and Zhang 2005; Kusdiana, and Saka 2001; Kusdiana and Saka 2004; Savage et al. 1995). This process requires a high alcohol to oil ratio (42:1 mole ratio). The reaction is complete in about 3 to 5 min. The process requires high pressure vessels that can be quite expensive.

The energy consumption also is higher than the conventional processes. The reaction must be quenched very rapidly so the products do not decompose. With regard to environmentally friendly aspects, biodiesel in supercritical methanol (SCM) does not require any catalysts and does not generate significant wastes. Moreover, this method is highly tolerant

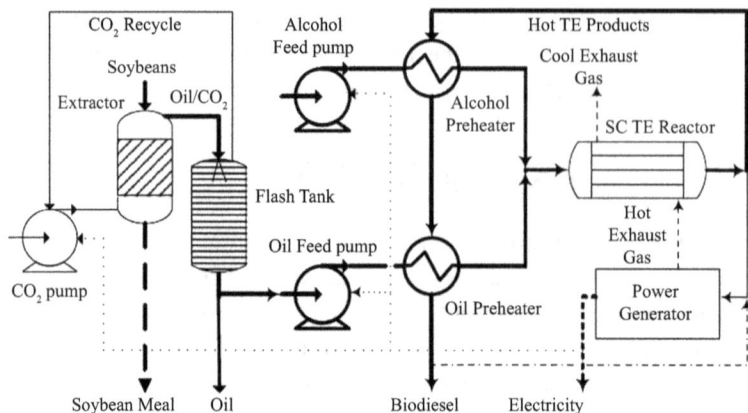

Figure 3.18. Supercritical technology system for biodiesel production.

against the presence of water in oils/fats, thus being applicable for various low-grade waste oils/fats. However, the usage of energy intensive synthesis process is the main disadvantage to push this technology from laboratory to industrial scale. Thus, future study should focus on the reduction of extreme operating parameters while maintaining the high conversion rate.

3.10 BIODIESEL STORAGE STABILITY

One of the main criteria for the quality of biodiesel is storage stability. Oxidation stability of biodiesel is an important issue because fatty acid derivatives are more sensitive to oxidative degradation than mineral fuel. The vegetable oil, fats, and their biodiesel suffer with the drawback of deterioration of its quality during long-term storage unlike petroleum diesel due to a large number of environmental and other factors making the fuel stability and quality questionable. There are various types of stabilities like oxidation, storage and thermal, playing key roles in making the fuel unstable. Vegetable oil derivatives especially tend to deteriorate owing to hydrolytic and oxidative reactions. Their degree of unsaturation makes them susceptible to thermal and/or oxidative polymerization, which may lead to the formation of insoluble products that cause problems within the fuel system, especially in the injection pump. The major fuel quality concerns as suggested by several researchers (Hoekman et al. 2012; Knothe 2008) are as follows:

- Stability and deposit formation
- Cold temperature handling and operability

- Solvency
- Microbial contaminants
- Water separation
- Material compatibility

Fatty acid alkyl chains have varying numbers of double bonds. Generally, the rate of oxidation of fatty acid alkyl esters depends on the number of double bonds and their position on the chain (Bouaid, Martinez, and Aracil 2007). When multiple double bonds are present, they are in allylic position, which means they are separated by a single methylene group. Common FAME such as rapeseed methyl ester (RME), canola methyl ester (CME), soybean methyl ester (SME), and tallow methyl ester (TME) contain primarily C16 to C18 carbon chains with zero to three double bonds. The 28-carbon chain oleic acid contains one double bond, two for linoleic acid and three for the linoleic acid. The relative oxidation rates for these C18 esters are linolenic > linoleic >> oleic (Cosgrove, Church, and Pryor 1987). Several studies related to the stability of bio-diesel have been reported in the literature. Westbrook (2005) has examined the storage stability of the B100 by the ASTM D4625 for a 12-week period. The author reported wide variations of insolubles formation, acid number, and viscosity increase. The least stable samples exhibited unacceptable levels of insolubles and acidity four to eight weeks into the test. McCormick et al. (2007) examined the stability characteristics of bio-diesel samples that were commercially available at blenders and distributors during 2004 and showed that the stability range results primarily from differences in fatty acid makeup and natural antioxidant content. Dunn (2008) studied the deterioration of RME under different storage conditions, including changes in acidity, peroxide value, and viscosity, and found that acid value, peroxide value, and viscosity increased with time. Several studies have showed that antioxidants improve bio-diesel oxidation stability. Plant-derived bio-diesel contains naturally occurring tocopherols, which slightly stabilize the bio-diesel (Serrano et al. 2013). Distillation of bio-diesel removes the natural antioxidant and reduces the Rancimat oxidation stability. Schneller et al. (2013) have shown that the addition of antioxidants can improve bio-diesel oxidation stability. The presence of high levels of oxidation products in the bio-diesel can lead to the formation of insoluble gums and sediment deposits in the fuel systems that can influence vehicle operability. This is one of the main concerns for engine and fuel injector manufacturers. Terry, McCormick, and Natarajan (2006) showed that at very high levels of oxidation, bio-diesel blends can separate into two phases to cause fuel pump and injector operational problems or lacquer deposits on fuel system components.

The major economic factor to consider for input costs of biodiesel production is the feedstock, which is about 80 percent of the total operating cost. Other important costs are labor, methanol, and catalyst, which must be added to the feedstock. Using an estimated process cost, exclusive of feedstock cost, of US$0.158/l ($0.60/gal) for biodiesel production, and estimating a feedstock cost of US$0.539/l ($2.04/gal) for refined soy oil, an overall cost of US$0.70/l ($2.64/gal) for the production of soy-based biodiesel was estimated (Hoekman et al. 2012). Even though the portion related to the processing and conversion of the feedstock in the total product cost is often negligible, the role of the production technology is not to be underestimated. In fact, specific technological approaches are required to treat multiple feedstock, including low-cost ones like waste oils. In addition, in the prospective of further biodiesel market development there is a tendency of equilibration of feedstock prices that will bring about a tight competition between production know-how.

Homogenous catalysts can be replaced by heterogeneous catalysts in order to run the transesterification process in continuous mode and avoid unfavorable saponification phenomena and related product-separation issues, as well as to allow other simple downstream processing steps and recovery of the catalyst. Overall, heterogeneous catalysts are more environmentally friendly than homogeneous ones and represent a more sustainable mode of resource management. However, there are still improvements to be made on the formulations as to allow their higher intrinsic efficiency in this reaction. Generally, heterogeneous catalysts are less active and suffer from deactivation phenomena. Enzymes can be called new-generation catalysts for the production of first-generation biodiesel from vegetable oils, as they offer a further set of improvements to the traditional process. Enzymes such as lipases promote the energetically favorable low-temperature transesterification process with yet lower environmental impact. As in the case of heterogeneous catalysts, the saponification and related separation issues are avoided, but, additionally, both the transesterification of triglycerides and esterification of FFAs can be run in one pot and at a lower methanol to oil ratio. Given ample feedstock availability and slightly modified catalyst formulations and reactor design to address more acidic feedstock, the process can be cost-competitive. In fact, industrial-scale facilities are now in operation and a huge interest in related research indicates a promising future for biodiesel technology.

CHAPTER 4

METHODS FOR QUALITY ASSESSMENT OF BIODIESEL

Various analytical methods were developed for analyzing mixtures containing fatty acid esters and mono-, di-, and tri-glycerides obtained by the transesterification of vegetable oils. Figure 4.1 shows the transesterification reaction of these triacylglycerides (TAGs) with alcohol to obtain the most common fatty esters contained in biodiesel: palmitic (16:0), stearic (18:0), oleic (18:1, *cis*-9), linoleic (18:2, *cis*-9,12), and linolenic (18:3, *cis*-9,12,15) (Carvalho et al. 2012). The diversity of carbon chains, degree of unsaturation, stereochemistry (*cis/trans*), and position of double bonds in the carbon chain make biodiesel as a complex mixture that contains a broad spectrum of fatty acid types, complicating their characterization. Besides, the partial glycerols, unreacted triacylglycerols, unseparated glycerol, free fatty acids, residual alcohol, and catalyst can contaminate the final product (Knothe 2001). The contaminants can lead to severe operational problems when using biodiesel, such as engine deposits, filter clogging, or fuel deterioration. In the United States, a provisional American Society for Testing and Materials (ASTM) standard PS 121 has been established (Howell 1997). In some European countries, such as Austria, the Czech Republic, France, Germany, and Italy, standards have been developed that limit the amount of contaminants in biodiesel fuel. In these standards, restrictions are placed on the individual contaminants by inclusion of items such as free and total glycerol for limiting glycerol and acylglycerols, flash point for limiting residual alcohol, acid value for limiting free fatty acids, and ash value for limiting residual catalyst. The determination of biodiesel fuel quality is therefore an issue of great importance to the successful commercialization of this fuel. Continuously high fuel quality with no operational problems is a prerequisite for market acceptance of biodiesel.

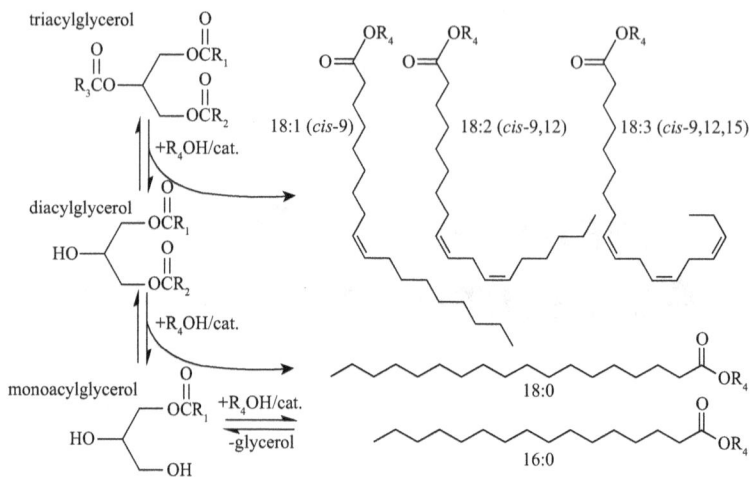

Figure 4.1. Transesterification of triacylglycerol (any feed stock). R_1, R_2, R_3 are the chains of aliphatic ester and R_4 is of alcohols. The fatty acid esters palmitic are (16:1), estearic (18:0), oleic (18:1), linoleic (18:2), and linolenic (18:3).

The ideal analytical method for a product such as biodiesel would be able to reliably quantify all contaminants even at trace levels. The fatty acid profiles are of considerable importance in the biodiesel analysis. Some chromatographic methods have been created and improved to analyze the biodiesel. Among these, thin layer chromatography (TLC), gas chromatography (GC), high-performance liquid chromatography (HPLC), gel permeation chromatography (GPC), nuclear magnetic resonance (NMR), and near-infrared spectroscopy (NIR) are included. The main reasons are likely the higher equipment costs and the higher investment in technical skills of personnel needed to interpret the data. It is important to note that, in order to satisfy the requirements of biodiesel standards, the quantification of individual compounds in biodiesel is not necessary but the quantification of classes of compounds. For example, for the determination of total glycerol, it does not matter which monoacylglycerol (e.g., monolein or monostearin) the glycerol stems from. The methods for detection of biodiesel quality will be described in subsequent sections.

4.1 GAS CHROMATOGRAPHY METHOD

A gas chromatographic method for the simultaneous determination of glycerol, mono-, di-, and tri-glycerides in vegetable oil methyl esters (ME) has been developed. Biodiesel standards have been established or are being developed in various countries and regions around the world (De Jong and Suijker 2012; Knothe 2006) including the following:

Method EN 14103—Determination of ester and linolenic ester ME content in fatty acid methyl ester (FAME).

Method EN 14105—Determination of free and total glycerol and mono-, di-, and tri-glyceride content in FAME.

Method EN 14110—Determination of methanol content in FAME.

Method ASTM D6584-10a—Determination of free and total glycerin content in B-100 ME.

Method EN 15779:2012—Determination of polyunsaturated fatty acid methyl esters.

Method ASTM D6584-10a—Determination of free and total glycerin content in B-100 ME.

Method EN 15779—Determination of polyunsaturated FAME.

The important parameter to assess the biodiesel quality are the carrier gas, internal standard, solvent, column dimension, injection volume, injection temperature, oven temperature (initial-hold-final), ramping of temperature inside the oven and flame ionization detector (FID) temperature (final temperature and holding time). It was reported (Ragonese et al. 2009) that the high polarity of the ionic liquid stationary phase allowed the separation of the FAMEs, from the less-retained hydrocarbons, thus avoiding the requirement of a hydrocarbon liquid chromatography pre-separation.

Ragonese et al. (2009) compared with the results derived from the analyses of a soybean FAMEs B20 sample, carried out on an SLB-IL100 conventional column (30 m × 0.25 mm i.d. × 0.20 mm d_f), and compared with those attained on a polyethylene glycol column, of equivalent dimensions. Conventional and fast GC methods, for the analysis of FAMEs in diesel blends, were developed on an SLB-IL100 30 m × 0.25 mm × 0.20 μm and on an SLB-IL100 12 m × 0.10 mm i.d. × 0.08 μm d_f column, respectively (KianHee, Yasir, and Kudumpor 2012; Seeley et al. 2007). The optimized IL methods were subjected to validation: Retention time and peak area intra-day precision were good. In principle, glycerol, mono-, di-, and tri-glycerides can be analyzed on highly inert columns coated with polar stationary phases without derivatization. The inertness of the column, required to obtain good peak shapes and satisfactory recovery, cannot be easily maintained in routine analysis (Meher, Vidya Sagar, and Naik 2006). Trimethylsilylation of the free hydroxyl groups of glycerol, mono-, and di-glycerides, however, ensures excellent peak shapes, good recoveries, and low detection limits and enormously improves the ruggedness of the procedure.

In a typical procedure (Di Serio et al. 2006; Islam et al. 2013a, 2013b), the fatty acid composition was determined by GC using an internal standard. The samples were analyzed by a Shimadzu GC-14B gas chromatograph equipped with a FID and a capillary column Rtx-65 (30 m × 0.5 mm × 0.25 μm). The injector and detector temperatures were 240°C and 280°C, respectively. Methyl heptadecanoate (0.1 g) was dissolved in 100 mL of hexane to prepare the standard solution. In all, 0.2 g of biodiesel sample was dissolved in 10 mL of standard solution for GC analysis and a 0.5 microliter sample was injected into the GC, which uses helium as a carrier gas. The content in FAME yield was determined in accordance with European regulated procedure EN 14103 (Bolognini et al. 2002).

To calculate the biodiesel yield, the response factors, R_f, for each compound have been calculated using the correspondent standard compound according to Equation 4.1.

$$R_f = \left(\frac{A_{is}}{A_{rs}}\right) \times \left(\frac{C_{rs}}{C_{is}}\right) \qquad (4.1)$$

where,

A_{is} = area of internal standard
C_{is} = concentration of internal standard
A_{rs} = *area of standard references*
C_{rs} = concentration of standard references

The ME was calculated using Equation 4.2:

$$ME = \frac{C_{iss} \times A_{if} \times R_f}{A_{iss}} \qquad (4.2)$$

where,

C_{iss} = concentration of internal standard in the sample
A_{iss} = area of internal standard in the sample
A_{if} = area of individual FAMEs compound in the sample

The biodiesel yield (%) was according to Equation 4.3:

$$\text{Biodiesel yield (\%)} = \frac{\text{total amount of ME (mol)}}{3 \times \text{charged amount of triglycerols (mol)}} \times 100 \quad (4.3)$$

Example of calculation of FAME yield (%) using GC according to EN 14103

1. Determination of response factor (RF) using reference standards
The fatty acid methyl ester in biodiesel sample was identified and quantified by comparing their retention times and peak areas to those of standard sample. Table 4.1 shows that the outcome in chromatograph of the compounds is in the following sequence: methyl myristate, methyl palmitate, internal standard, methyl stearate, methyl oleate, and methyl linoleate. Figure 4.2 shows retention time and surface area of (a) standard ME and (b) experimental biodiesel product in palm oil.

The RFs of each fatty acid standard are calculated using Equation 4.4:

$$RF_{RS} = \frac{A_{is} \times C_{RS}}{A_{RS} \times C_{is}} \tag{4.4}$$

where,

A_{is} = Area of internal standard
C_{RS} = Concentration of reference standard in solution (ppm) (1000 ppm)
A_{RS} = Area of reference standard
C_{is} = Concentration of internal standard (methyl hepta dodecanoate) in reference standard solution (ppm) (1000 ppm)

2. Calculation of FAMEs yield (%) from the sample (sample $NaNO_3$ $0.30_{g/g\ support}$)

$$A_{ppm} = C_{iss} \times A_{FCS} \times RF_{RS}/A_{iss} \tag{4.5}$$

C_{iss} = Concentration of internal standard in the sample (ppm) (1000 ppm used in this experiment)
A_{iss} = Area of internal standard (methyl hepta dodecanoate) in the sample (from Table 4.2)
A_{FCS} = Area of individual FAMEs compound in the sample (from Table 4.2)
RF_{RS} = Response factor of the respective reference standard (from Table 4.1)

$$B_{mg} = \frac{A_{ppm} \times 10}{1000} \tag{4.6}$$

(Volume of biodiesel mixed solution was 10 mL, ppm = mg/1000 mL)

$$c_g \text{ of FAMEs compound in the sample} = \frac{\left(B_{mg} \times \dfrac{12.95}{0.2}\right)}{1000} \tag{4.7}$$

Table 4.1. Retention time and surface area for the reference standards

FAMEs standards	Run 1		Run 2		Run 3		Avg. area	Response factor (RF)
	Retention time	Area	Retention time	Area	Retention time	Area		
Methyl hepta dodecanoate (internal standard)	10.353	12,146	10.346	12,900	10.336	11,949	12,331.66667	—
Methyl myristate (mm)	8.394	10,280	8.389	10,398	8.38	9,625	10,101	1.220836
Methyl palmitate (mp)	9.654	9,816	9.617	10,322	9.637	9,672	9,936.666667	1.241027
Methyl stearate (ms)	11.131	9,159	11.123	9,792	11.111	9,074	9,341.666667	1.320071
Methyl oliate	11.223	10,036	11.214	1,0626	11.201	9,989	10,217	1.206975
Methyl linoleate	11.429	6,891	11.419	7,296	11.406	6,996	7,061	1.746448

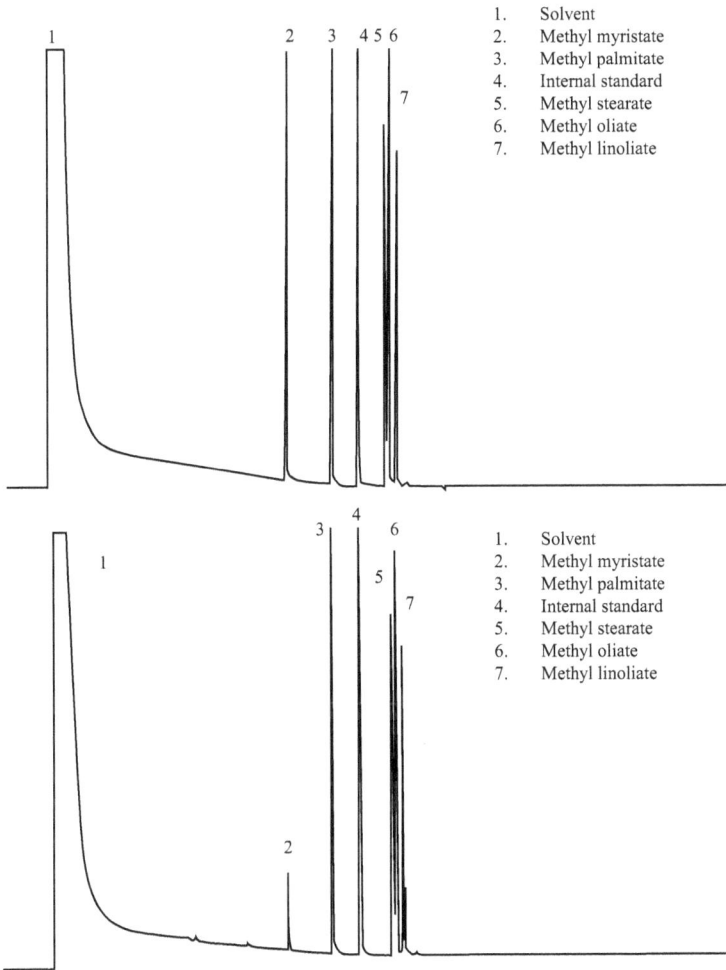

1.	Solvent
2.	Methyl myristate
3.	Methyl palmitate
4.	Internal standard
5.	Methyl stearate
6.	Methyl oliate
7.	Methyl linoliate

1.	Solvent
2.	Methyl myristate
3.	Methyl palmitate
4.	Internal standard
5.	Methyl stearate
6.	Methyl oliate
7.	Methyl linoliate

Figure 4.2. Gas chromatogram of (a) standard ME and (b) experimental biodiesel product in palm oil (Islam et al. 2013b).

where,

12.95 g of biodiesel obtained after reactions

0.2 g biodiesel mixed with 10 mL of mixed (hexane and internal standard) solution

$$^{D}Mol\ (ME) = \frac{\text{g of FAMEs compound in the sample}}{\text{Molecular weight of individual ME}} \quad (4.8)$$

$$^{E}FAME\ Yield\ (\%) = \frac{\text{Amount experimentally FAME produced (mole)}}{\text{Amount of FAME produced theoretically (mole)}} \times 100 \quad (4.9)$$

Table 4.2. Calculation of different parameters using response factor

Standards	Retention time	Area	[A]ppm	[B]mg	[C]g of FAMEs compound in the sample	[D]Mol (ME)	Total methyl ester	[E]Total yield (%)
Internal standard	10.363	11,311	—	—	—	—	—	—
Methyl myristate	8.383	1,977	213.3845998	2.133845998	0.138166528	0.000570017	0.046182275	87.30108716
Methyl palmitate	9.733	74,770	8,203.655865	82.03655865	5.311867173	0.019640847		
Methyl stearate	11.325	86,289	10,070.51879	100.7051879	6.520660915	0.021844028		
Methyl oliate	11.5	15,752	1,680.865968	16.80865968	1.088360714	0.003670756		
Methyl linoleate	11.74	1,345	207.6714749	2.076714749	0.13446728	0.000456626		

$$\text{Amount of FAME produced theoretically (mole)} = \frac{\text{Amount of oil used for reaction}}{\text{Molicular weight of palm oil}} \times 3 \quad (4.10)$$

where,

Initial amount of palm oil used for reaction = 15 g
Molecular weight of palm oil = 849.5 g/mol

The molecular weight of palm oil was calculated using the molecular weight of different standards from Table 4.3.

Conversion of vegetable oil to biodiesel is usually monitored by GC. Several researchers (Rezanka and Sigler 2007; Fedosov, Fernandes, and Firdaus 2014) have reported that the GC is not always convenient due to the following reasons: (i) an elaborate derivatization of the samples, (ii) inhibition of this process by methanol and water, and (iii) low stability of the derivatives under storage. Thus, many researchers have been directed toward the use of HPLC methods for monitoring the conversion or yield of biodiesel.

4.2 HPLC METHOD

HPLC offers a useful alternative to GC and many liquid chromatographic methods have been developed. Many workers (Adlof, Copes, and Emken 1995; Carvalho et al. 2012; Freedman et al. 1986a, 1986b), however, may find that HPLC offers some advantage and indicates some critical points of GC: (i) the presence of heat-labile compounds affects the quantification of FAME; (ii) the carbon chain polyunsaturated of the fatty acids may undergo structural changes, isomerization, and decomposition under high temperatures; (iii) it is not possible to collect fractions of the separated fatty acid esters for further analysis; (iv) baseline drift; and (v) GC analysis frequently requires derivatization step by saponification furthered

Table 4.3. Molecular weight of different standards

Name of standards	Molecular weight (g/mol)
Methyl myristate	242.39
Methyl palmitate	270.45
Methyl stearate	298.51
Methyl oliate	296.49
Methyl linoleate	294.84

to methylation, consuming reagents as hexane, BF_3, and NaCl, being time-consuming and labor-intensive (Freedman et al. 1986; Nollet and Toldrá 2012).

The important parameters to assess the biodiesel quality using HPLC are the column dimension, injection volume, and flow rate with gradient elution of solvent. HPLC with pulsed amperometric detection (the detection limit is usually 10 to 100 times lower than for amperometric detection and the detection limit is 1 mg/g) was used to determine the amount of free glycerol in vegetable oil esters (Garba, Alhassan, and Abdulsalami 2006; Meher, Vidya Sagar, and Naik 2006). Lozano et al. (1996) reported that the HPLC-PAD method was more simple, rapid, and accurate to determine the free glycerol in vegetable oil esters due to its high sensitivity.

Albuquerque et al. (2008) determined the composition of the trans-esterification products derived from the methanolysis of sunflower oil by HPLC equipped with multiwavelength detector (MD-2015) operated at 30°C with 0.7 mL/min flow rate of 80 percent acetonitrile solution containing 20 percent of 0.1 percent H_3PO_4 as a mobile phase. Several researchers (Berchmans and Hirata 2008; Gaita 2006; Joshi, Toler, and Walker 2008) used evaporative light scattering detector (ELSD) detector to qualitatively and quantitatively analyze the conversion of biodiesel. A HPLC system equipped with C18 column of length 150 mm and inner diameter 4.6 mm was used for analyses (Joshi, Toler, and Walker 2008; Myller et al. 2012). The mobile phase was a mixture of acetonitrile and dichloromethane, with a gradient of dichloromethane maintained to separate the biodiesel sample. Joshi, Toler, and Walker (2008) maintained gradient time—(0, 15, 30, 32, 35) min—and percentage dichloromethane—(0, 15, 70, 70, 0). Hundred percent of acetonitrile was also used as a mobile phase with a flow rate of 1.0 mL/min at the detection wavelength of 300 nm (Carvalho et al. 2012). The sample volume was 20 µL and a peak identification was made by comparing the reaction time between the sample and the standard compound (Joshi, Toler, and Walker 2008). The biodiesel sample could be dissolved with 2-propanol-hexane at a ratio of 5:4 (v/v) or 1:15 dilution of biodiesel in dichloromethane (Albuquerque et al. 2008; XiaoHu, Xi, and Feng 2011). The FAME peak identification was determined by the comparison of retention time of the reference standards in the same condition (Carvalho et al. 2012; Monteiro et al. 2008; Scholfield 1975), as shown in Figure 4.3a.

In relation to instrument calibration, the aim of linear regression is to establish the equation that best describes the linear relationship between instrument response (y) and analyte level (x). The relationship is described by the equation of the line, that is, $y = mx + c$, where m is the gradient of the line and c is its intercept with the y-axis. Linear regression establishes

(a)

Elution Order:
1. Glycerol
2. Butanetriol (istd #1)
3. C16 Methyl Ester
4. C18 Methyl Ester
5. C20 Methyl Ester
6. Monoleins
7. Tricaprin (istd #2)
8. Dioleins
9. Triolein

Time, min.

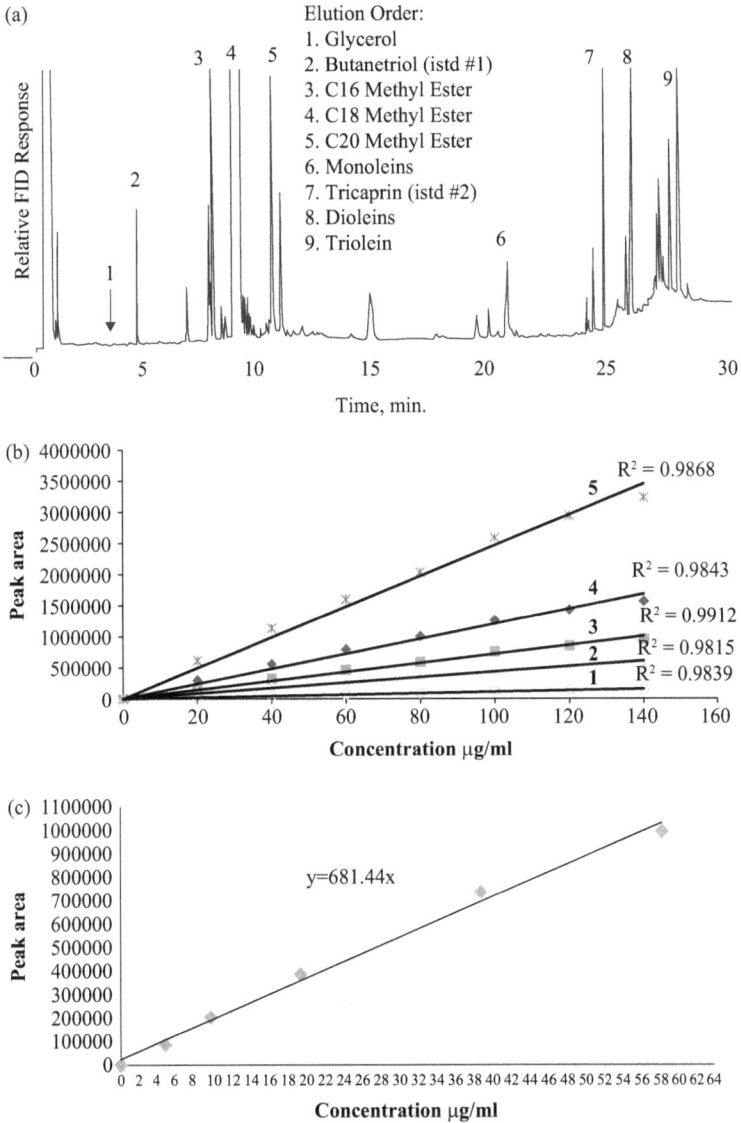

Figure 4.3. (a) HPLC chromatogram of standards; (b) calibration curves for (1) methyl myristate, (2) methyl palmitate, (3) methyl stearate, (4) methyl oliate; (5) methyl linoleate; and (c) calibration curve for methyloleate.

the values of m and c, which best describes the relationship between the data sets. This is a reasonable assumption for many analytical methods as it is possible to prepare standards where the uncertainty in the concentration is insignificant compared with the random variability of the analytical instrument. It is therefore essential to ensure that the instrument response data and the standard on concentrations are correctly assigned. Many researchers (Carvalho et al. 2012; Monteiro et al. 2008; Scholfield 1975) have reported to determine the conversion of biodiesel using the HPLC result plotted with linear regression lines, as shown in Figure 4.3b and Figure 4.3c. The example of standards calibration curve for the methyloleate (Carvalho et al. 2012; Murphy 2012) is shown in Figure 4.3c. The curve obtained from the HPLC for the methyloleate counts 8.0×10^5 and the original mass of sample was 58 mg. Form the linear regression lines,

$$Y = mx + C$$
$$8.0 \times 10^5 = 681.44x$$
$$X = 1173.98 \ \mu g/mL$$

Initially 10 mL sample was prepared for analysis. So, the mass of methyloleate is

$$X = 11739.8 \ \mu g$$

It can be calculated as the percentage of methyloleate by the following equation:

$$\text{Percentage of monoglycerides} = \frac{11739.8 \ \mu g}{58000 \ \mu g} \times 100 = 20\%$$

4.3 GEL PERMEATION CHROMATOGRAPHY METHOD

A method for simultaneous analysis of transesterification reaction products-monoglycerides, diglycerides, triglycerides (TG), glycerol, and ME was developed using GPC coupled with refractive index detector (Darnoko, Cheryan, and Perkins 2000). The mobile phase was HPLC grade tetrahydrofuran at a flow rate of 0.5 mL/min at room temperature and the sample injection size was 10 mL. Sample preparation involves only dilution and neutralization. For analysis, 300 mg of the sample was taken from transesterification reactor and neutralized by adding 5 mL HPLC grade tetrahydrofuran and one drop of 0.6 N HCl. The samples were then kept at 20°C until analysis. Reproducibility of the method was good: Analysis of palm oil transesterification products at different levels of conversion showed a relative standard deviation of 0.27 to 3.87 percent. Similarly, GPC was used to evaluate the influence of different variables affecting the transesterification of rapeseed oil with anhydrous ethanol and sodium ethoxide as catalyst (Fillières, Benjelloun-Mlayah, and Delmas 1995; Meher, Vidya Sagar, and Naik 2006). GPC has made the quantitation of ethyl esters, mono-, di-, and tri-glycerides and glycerol possible.

4.4   TLC METHOD

A method for assaying glycerol in biodiesel product, both the separated layers were developed using TLC (Fontana et al. 2009). The stationary phase for TLC realization was Merck SG-60 chromatoplate (Whitehouse Station, NJ) and the mobile phase was a mixture of toluene–chloroform–acetone (7:2:1, v/v/v). Butanol and water (1:1) was used by Cai (2014) as mobile phase and the plate was stained with 0.5 percent $KMnO_4$ (dissolved in 1 N NaOH) to view the spots. Sample preparation involves the normalization of biodiesel and acylglycerol standards (monoolein, monoacylglyceride (MAG); diolein isomers, diacylglycerides (DAGs); triolein, TAG) to 50mg/mL in isopropanol. One or two microliters were then applied to the chromatoplate. To develop the color through a fine spray, 1 percent hot p-anisaldehyde in methanol–sulfuric acid (9:1, v/v) was used. For the development of rose-violet color, p-anisaldehyde acid sprayed on the TLC plate was warmed for 3 to 5 min in a hot Cimarec plate and the pictures were then taken with a Sony Cyber-shot 5 megapixel camera (Figure 4.4). Similarly, a mixture of 0.5 mL chloroform and methanol (9:1) was dissolved in one drop of the esterification products (Fattah et al. 2014). A drop from the final mixture was then applied over the TLC paper immersed into a 100-mL beaker containing 10 mL of a mixture of chloroform and methanol (9:1), which allows a contact between the edges of the TLC paper with the solvent mixture. The TLC paper was allowed to stand until the solvent reached a level that must be just below the end line. The TLC paper was then dried and the bands of the different glycerides produced were detected (Figure 4.4).

4.5   SPECTROSCOPIC METHODS

Spectroscopic methods also have been reported for the analysis of biodiesel and/or monitoring of the transesterification reaction. Proton nuclear magnetic resonance (^{1}H NMR) spectroscopy has been used extensively to analyze biodiesel, vegetable oil feeds, reaction intermediates, and final

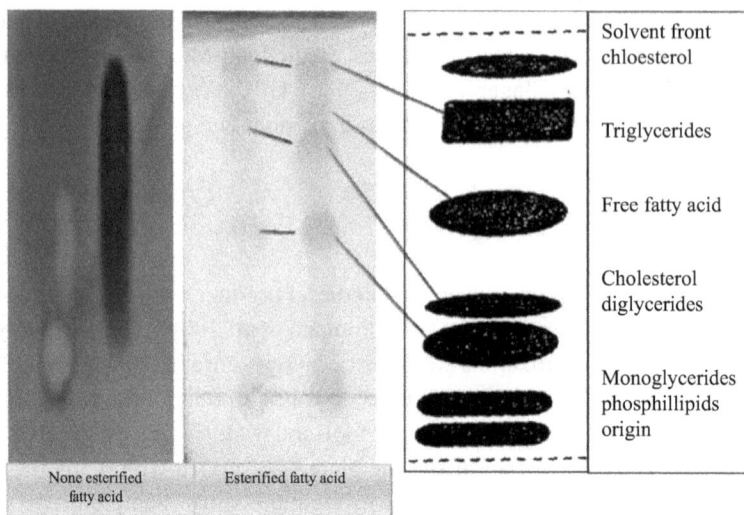

Figure 4.4. Thin layer chromatograph of biodiesel.

products of the biodiesel transesterification process. Proton NMR provides a good probe for biodiesel since ^{1}H is the most naturally abundant and most sensitive NMR active isotope. Relatively narrow line widths of a few Hertz are obtained for ^{1}H spectra so that magnetically unique nuclei are resolved at many field strengths.

4.5.1   NUCLEAR MAGNETIC RESONANCE

The first report on spectroscopic determination of the yield of transesterification reaction utilized ^{1}H NMR depicting its progressing spectrum (Gelbard et al. 1995). Knothe and Kenar have shown that integrals of resonances in ^{1}H spectra can be used to determine the relative amounts of fatty acids in vegetable oils and ME mixtures when the source of the oil feedstock is known (Knothe and Kenar 2004). Previous work by Diehl and Randel (2007) has shown the ability of NMR to quantify blends of biodiesel and petroleum diesel. Diehl and Randel (2007) presented ^{1}H NMR spectroscopy as a routine analysis for diesel mixed with biodiesel and gasoline. For sample preparation, the test sample of 500 mg and 10 mg of internal standard (chloroform-d, $CDCl_3$) was dissolved in 200 µL benzene-d_6 (Diehl et al. 2007). ^{1}H NMR has been used to monitor the transesterification reaction of *Jatropha* oil (Kapilan 2012), see for example Figure 4.5.

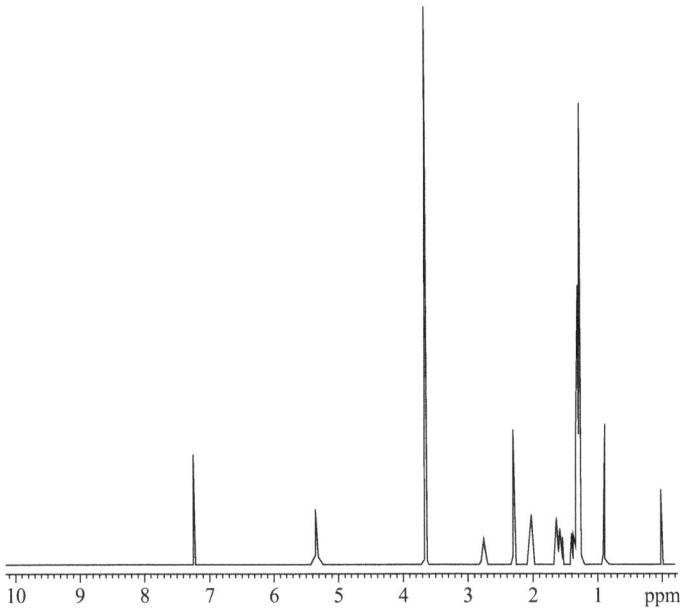

Figure 4.5. NMR spectra of biodiesel.

In the NMR spectra, TG protons on acyl groups resonate at 0.8 to 2.9 ppm, while protons H-1, H-2, and H-3, appear at a downfield of 4.0 to 5.6 ppm. When one or two acyl groups migrate from TGs, H-1, H-2, and H-3 shift toward a higher field. Compared to the NMR spectrum of *Jatropha* oil, a large signal at 3.6 pm was observed, which was assigned to the methyl protons of the esters. In addition, some new peaks appeared. This was attributed to the methanolysis products of mono and di-glycerides. An equation was given by the authors (Jin et al. 2007) as follows:

$$C = 100 \times \frac{2A_{ME}}{3A_{-CH2}} \tag{4.11}$$

where,

C conversion of triacylglycerol feedstock (vegetable oil) to the corresponding ME.

A_{ME} integration value of the protons of the ME (the strong signal peak)

A_{CH2} integration value of the methylene protons

Factors 2 and 3 derive from the fact that the methylene carbon possesses two points and the alcohol (methylene-derived) carbon has three attached protons.

4.5.2 NEAR-INFRARED SPECTROSCOPY

NIR spectroscopy was used to monitor the transesterification reaction (Knothe 1999). NIR spectra were obtained with the aid of a fiber-optic probe coupled to the spectrometer, which renders their acquisition particularly easy and time-efficient. The absorption bands of TG and their corresponding FAMEs are close to each other in the mid-IR spectral range, but are quite apart in near-IR spectral range (above 4000 cm^{-1}) (Holman and Edmondson 1956).

Knothe (1999, 2001) reported that the methyl esters display IR absorption bands in both the 4425 to 4430 cm^{-1} and 6005 cm^{-1} regions while the TG (esters of triglycol) in vegetable oils show only a shoulder band in the 4425 to 4430 cm^{-1} range, as shown in Figure 4.6. The accuracy of the NIR method in distinguishing triacylglycerols and methyl esters is in the range of 1 to 1.5 percent, although in most cases better results are achieved. To circumvent this difficulty, an inductive method can be applied (Knothe

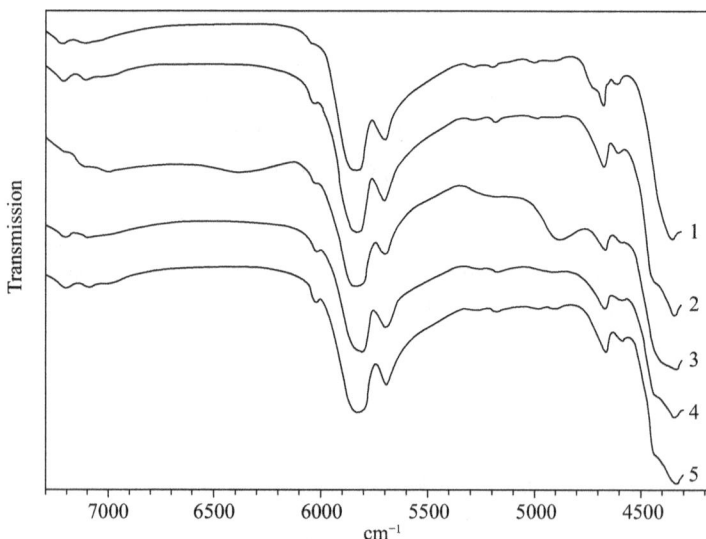

Figure 4.6. Near-infrared spectra in the region 7300–4300 cm^{-1} of (1) soybean oil (feedstock); (2) methyl soyate; (3) methyl soyate contaminated with methanol (80 mL methyl soyate, 20 mL methanol corresponding to 81.49 wt% methyl soyate and 18.51 wt% methanol); (4) methyl soyate contaminated with glycerol; and (5) methyl soyate contaminated with free fatty acid. The wavenumbers of some salient peaks useful for quantitation as discussed in the text are noted.

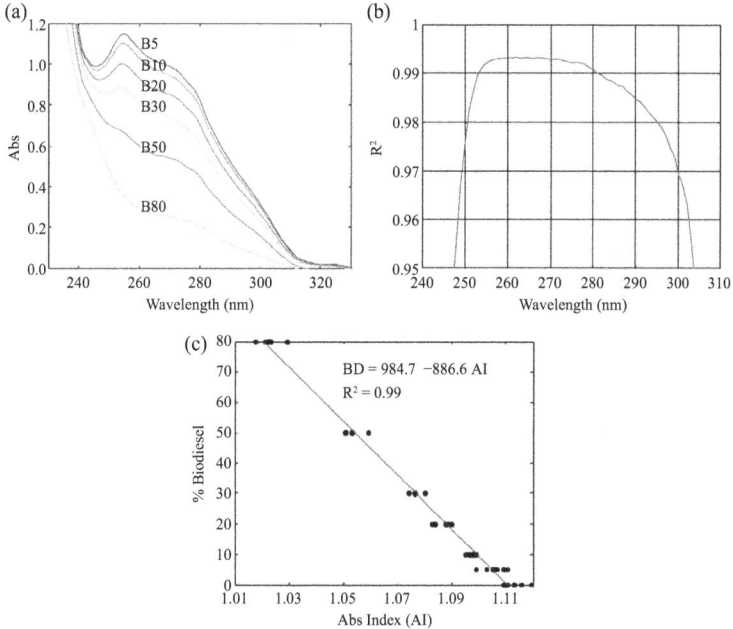

Figure 4.7. (a) UV absorbance spectra of soy methyl ester; (b) R^2 of linear relation of biodiesel percentage and absorbance a function of wavelength; and (c) transformed relation between blend level and absorbance index.

2001). The inductive method consists of verifying by GC that a biodiesel sample meets prescribed biodiesel standards.

4.5.3 ULTRAVIOLET ABSORPTION SPECTRA

Determining the relative amounts of biodiesel and diesel in the blend from the absorbance was reported by Artur and Shrestha (2007). A single-beam general-purpose spectrophotometer was used to determine the UV absorption spectra of the biodiesel samples and the biodiesel blends with diesel (Zawadzki, Shrestha, and He 2007). The UV absorbance spectrum of a progressing transesterification reaction is reported by Zawadzki, Shrestha, and He (2007) as shown in Figure 4.7a.

In order to find the wavelength for the best correlation, R^2 was calculated by fitting a linear line for each wavelength from 245 to 305 nm and plotted against the corresponding wavelengths (Figure 4.7b). Therefore, the absorbencies at three wavelengths where the aromatics were best absorbed (265, 273, and 280 nm) were chosen to extract the absorbance index. The absorbance index is a measure of the shape of the absorbance curve and is defined as follows:

$$AI = 10^{\left(A_{273e} - \frac{A_{265} + A_{280}}{2}\right)} \tag{4.12}$$

where, AI is the absorbance index, and A_{xxx} is the absorbance at the xxx nm wavelength. The AI was found to be linearly correlated with the blend level. The values of AI were calculated at various blend levels from B5 to B80 using the equation. Finally, linear Equation 2 obtained from the Figure 4.7c was used for the determination of the biodiesel blend (BD).

$$BD = 984.7 - 886.6 \, AI \tag{4.13}$$

The method was found to be applicable to any biodiesel feedstock and independent of the diesel fuel origins included in this research.

4.6 OTHER METHODS

4.6.1 VISCOMETRY

The viscosity of a fluid can be expressed as a "dynamic viscosity" and a "kinematic viscosity." Dynamic viscosity is measured in units called "centipoise." Kinematic viscosity takes into account the fluid density and is measured in units called "centistokes." It is also important when comparing values to note the temperature the measurements were taken at. In general, the viscosity of a liquid will be reduced as the temperature rises. For example, the range of viscosity seen for rapeseed methyl-ester in the chart is 4.43 to 6.7 centistokes at 40°C; this drops to around 2.4 centistokes at 100°C (Ma and Hanna 1999). Viscosities determined at 20°C and 37.8°C were in good agreement with GC analyses conducted for verification purposes. The viscometric method, especially results obtained at 20°C, is reported to be suitable for process-control purposes due to its rapidity (De Filippis et al. 1995). The high viscosity of the vegetable oils was the cause of severe operational problems, such as engine deposits (Bruwer et al. 1981; Dunn, Knothe, and Bagby 1997; Knothe, Dunn, and Bagby 1997). This is a major reason why neat vegetable oils largely have been abandoned as alternative diesel fuels in favor of mono-alkyl esters such as methyl esters.

4.6.2 DETERMINATION OF FREE FATTY ACID PERCENTAGE AND ACID VALUE

A number of researchers have worked with feedstock that has elevated free fatty acid (FFA) levels (Freedman, Pryde, and Mounts 1984; Mittelbach and Tritthart 1988; Peterson et al. 1997; Wimmer 1995). However, in most cases, alkaline catalysts have been used and the FFAs were removed from the process stream as soap and considered waste. As FFA levels increase, this becomes undesirable because of the loss of feedstock as well as the deleterious effect of soap on glycerin separation. The soaps promote the formation of stable emulsions that prevent separation of the biodiesel from the glycerin during processing. Waste greases typically contain from 10 percent to 25 percent FFAs. This is far beyond the level that can be converted to biodiesel using an alkaline catalyst. An alternative process is to use acid catalysts that some researchers have claimed are more tolerant of free fatty acids (Canakci and Van Gerpen 2001; Aksoy et al. 1988; Liu 1994).

FFAs are the result of the breakdown of oil or biodiesel. FFA% is usually used to describe the FFA content of oils, while acid number (AN) is commonly used to describe the FFA content of finished biodiesel. With a little math, we can use the same titration procedure we use to titrate waste vegetable oil (WVO) to determine FFA% and AN.

Free Fatty Acids (%)

FFA% is the weight to weight ratio of FFA found in an oil sample. The weight of an oil sample divided into the weight of the FFA in that sample.

To calculate FFA% from a titration value, the formula is:

$$FFA\% = (v - b) \times N \times 28.2/w$$

v is the volume in mL of titration solution
b is the volume in mL of the blank
N is the normality of the titration solution
w is the weight of the sample of oil in grams

Since in the homebrew titration we do not record the blank, we set it to zero here.

For N we use 1 gram/L or 0.025N for NaOH
For w we use 1mL of oil, which typically weighs 0.92 grams
28.2 is the molecular weight of oleic acid divided by 10.

When we plug everything in we get the following:

$FFA\% = 0.766t$—for NaOH titrations
$FFA\% = 0.546t$—for KOH titrations
t is our titration results in milliliters.

Note that since we estimated both the density of our oil sample and the molecular weight of the average fatty acid in our oil, the result will also be an estimate. Even though it is an estimate, it is close enough for our use.

Acid Number

Acid number, normally measured in biodiesel rather than WVO, is the amount of KOH in milligram needed to react with the acid in an amount of oil in grams. The acid number is one of the ASTM tests for finished biodiesel.

Determination of acid number by potentiometric titration using ASTM D664.

Max value for finished biodiesel = 0.50 mg/g

$$AN = (v - b) \times N \times 56.1/w$$

v is the titration volume in mL
b is the blank in mL
N is the normality of the KOH solution
w is the weight of sample in grams

Since we are using 1g/L of KOH for our titration, N = 1/56.1.

Of course then N and 56.1 cancel each other out and our AN is now
$AN = (v - b)/w$ or
$AN = 1.08t$

where t is the results of a standard homebrew style KOH titration.

When performing potentiometric titration to determine the acid number, it is recommended to use an automatic potentiometric titrator in order to simplify running the test while minimizing human error. The titrant used in this test is 0.1 mol/L of KOH 2-propanol solution and the mixed solvent to be titrated is a 500:5:495 ratio of toluene, pure water, and 2-propanol, respectively. The soap content of biodiesel is another test that can be run using a colorimetric titration technique where the sample is titrated with a standard hydrochloric acid solution (in acetone) to the bromophenol blue endpoint. Redox titrations are based on oxidation-reduction reactions between the analyte and the titrant, indicated by changes in the electrical potential of the test solution. A biodiesel application of the redox titration is the determination of the iodine value, which is related to the degree of unsaturation (number of carbon double and triple bonds) of the sample, which, in turn, is another indicator of the fuel's stability (Bouaid, Martinez, and Aracil 2007; Ferrari, Oliveira, and Scabio 2005; Knothe 2007).

A higher iodine value number indicates a higher quantity of double bonds in the sample, with corresponding lower oxidation stability. The iodine value of biodiesel can vary significantly based on the feedstock used. For example, rapeseed methyl ester has an iodine value number of 97, compared to an iodine value number of 133 for soy methyl ester (Stoytcheva 2011). Many biodiesel fuel standards specify an upper limit for the iodine value of fuel in order to meet the specification. The downside to iodine value is the fact that the value does not take into account the positions of the double bonds. Therefore, the iodine value is not necessarily the best indicator of a fuel's oxidation stability.

Several analytical methods have been investigated for fuel quality assessment and production monitoring of biodiesel. The most intensively studied method is GC, while HPLC, NMR, and NIR have also been studied. GC is also the method used for verification that biodiesel meets prescribed standards due to its ability to detect low-level contaminants, although improvements to this method are possible. Physical property-based methods have been explored less, and it appears that this may be an area for further study. Although GC suffers from many drawbacks, it has been the most used technique for analyzing the complex mixture of compounds involved in the transesterification, including TAGs, DAG, MAG, mono-alkyl esters, alcohol, and free glycerol. It is worth mentioning that these determinations cannot be carried out in a unique analysis because different methods are required. However, no method can simultaneously satisfy all criteria of simultaneously determining all trace contaminants with minimal investment of time, cost, and labor. A fast and easy to use method that may be adaptable to production monitoring, such as NIR (or viscometry), can be used for routine analyses. For example, if measurements by NIR (or viscometry) at several turnover ratios indicate that the transesterification reaction is progressing as desired and that the NIR spectrum (viscosity) of the biodiesel product agrees with that of one known to meet biodiesel standards, then further, more complex, analyses would be unnecessary. Only if NIR (or viscometry) analyses indicate that there is a potential problem with the product would more complex and time-consuming analyses, for example, by GC, be warranted to determine the exact cause of the problem.

REFERENCES

Abdulla, R., and P. Ravindra. 2013. "Characterization of Cross Linked Burkholderia Cepacia Lipase in Alginate and κ-Carrageenan Hybrid Matrix." *Journal of the Taiwan Institute of Chemical Engineers* 44, no. 4, pp. 545–51. doi: 10.1016/j.jtice.2013.01.003.

ACEA. March 2009. "Biodiesel Guidelines." *European Automobile Manufacturers Association,* Brussels, Belgium, http://www.acea.be/images/uploads/files/20090423_B100_Guideline.pdf

Adlof, R.O., L.C. Copes, and E.A. Emken. 1995. "Analysis of the Monoenoic Fatty Acid Distribution in Hydrogenated Vegetable Oils by Silver-ion High-performance Liquid Chromatography." *Journal of the American Oil Chemists' Society* 72, no. 5, pp. 571–4. doi: 10.1007/BF02638858.

Agarwal, A.K., and L.M. Das. 2001. "Biodiesel Development and Characterization for Use as a Fuel in Compression Ignition Engines." *Journal of Engineering for Gas Turbines and Power* 123, no. 2, pp. 440–7. doi: 10.1115/1.1364522.

Akoh, C.C., S.W. Chang, G.C. Lee, and J.F. Shaw. 2007. "Enzymatic Approach to Biodiesel Production." *Journal of Agricultural and Food Chemistry* 55, no. 22, pp. 8995–9005. doi: 10.1021/jf071724y.

Aksoy, H.A., I. Kahraman, F. Karaosmanoglu, and H. Civelekoglu. 1988. "Evaluation of Turkish Sulphur Olive Oil as an Alternative Diesel Fuel." *Journal of the American Oil Chemists' Society* 65, no. 6, pp. 936–8. doi: 10.1007/BF02544514.

Alamu, O.J., M.A. Waheed, S.O. Jekayinfa, and T.A. Akintola. 2007. "Optimal Transesterification Duration for Biodiesel Production from Nigerian Palm Kernel Oil." *International Commission of Agricultural Engineering: The CIGR Ejournal. Manuscript EE 07 018* Vol. 9.

Albuquerque, M.C., J. Santamaría-González, J.M. Mérida-Robles, R. Moreno-Tost, E. Rodríguez-Castellón, A. Jiménez-López, and P. Maireles-Torres. 2008. "MgM (M= Al and Ca) Oxides As Basic Catalysts in Transesterification Processes." *Applied Catalysis A: General* 347, no. 2, pp. 162–8. doi: 10.1016/j.apcata.2008.06.016.

Antunes, W.M., C.D.O. Veloso, and C.A. Henriques. 2008. "Transesterification of Soybean Oil with Methanol Catalyzed by Basic Solids." *Catalysis Today* 133, pp. 548–54. doi: 10.1016/j.cattod.2007.12.055.

Armaroli, N., and V. Balzani. 2007. "The Future of Energy Supply: Challenges and Opportunities." *Angewandte Chemie International Edition* 46, no. 1–2, pp. 52–66. doi: 10.1002/anie.200602373.

Artur, Z., and D.S. Shrestha. 2009. Biodiesel/Diesel Blend Level Detection Using Absorbance. US Patent 20,090,316,139. Washington, DC: U.S. Patent and Trademark Office.

ASTM. 2002. "Standard Specification for Biodiesel Fuel (B100) Blend Stock for Distillate Fuels." *American Society for Testing and Materials*, D6751–02.

Atabani, A.E., A.S. Silitonga, H.C. Ong, T.M.I. Mahlia, H.H. Masjuki, I.A. Badruddin, and H. Fayaz. 2013. "Non-edible Vegetable Oils: A Critical Evaluation of Oil Extraction, Fatty Acid Compositions, Biodiesel Production, Characteristics, Engine Performance and Emissions Production." *Renewable and Sustainable Energy Reviews* 18, pp. 211–45. doi: 10.1016/j.rser.2012.10.013.

Atadashi, I.M., M.K. Aroua, and A.A. Aziz. 2011. "Biodiesel Separation and Purification: A Review." *Renewable Energy* 36, no. 2, pp. 437–43. doi: 10.1016/j.renene.2010.07.019.

Atadashi, I.M., M.K. Aroua, A.R. Abdul Aziz, and N.M.N. Sulaiman. 2011. "Membrane Biodiesel Production and Refining Technology: A Critical Review." *Renewable and Sustainable Energy Reviews* 15, no. 9, pp. 5051–62. doi: 10.1016/j.rser.2011.07.051.

Atadashi, I.M., M.K. Aroua, A.R. Abdul Aziz, and N.M.N. Sulaiman. 2012a. "High Quality Biodiesel Obtained Through Membrane Technology." *Journal of Membrane Science* 421, pp. 154–64. doi: 10.1016/j.memsci.2012.07.006.

Atadashi, I.M., M.K. Aroua, A.R. Abdul Aziz, and N.M.N. Sulaiman. 2012b. "The Effects of Water on Biodiesel Production and Refining Technologies: A Review." *Renewable and Sustainable Energy Reviews* 16, no. 5, pp. 3456–70. doi: 10.1016/j.rser.2012.03.004.

Bajpai, D., and V.K. Tyagi. 2006. "Biodiesel: Source, Production, Composition, Properties and Its Benefits." *Journal of Oleo Science* 55, no. 10, pp. 487–502. doi: 10.5650/jos.55.487.

Banga, S., P. Varshney, and N. Kumar, 2012. "Biodiesel Purification Using Organic Adsorbents: A Preliminary Study." *Journal of Biofuels* 3, no. 2, pp. 112–8. doi: 10.5958/j.0976-4763.3.2.012.

Barnard, T.M., N.E. Leadbeater, M.B. Boucher, L.M. Stencel, and B.A. Wilhite. 2007. "Continuous-flow Preparation of Biodiesel Using Microwave Heating." *Energy & Fuels* 21, no. 3, pp. 1777–81. doi: 10.1021/ef0606207.

Basinger, M., T. Reding, C. Williams, K.S. Lackner, and V. Modi. 2010. "Compression Ignition Engine Modifications for Straight Plant Oil Fueling in Remote Contexts: Modification Design and Short-run Testing. *Fuel* 89, no. 10, pp. 2925–38. doi: 10.1016/j.fuel.2010.04.028.

Behzadi, S., and M.M. Farid, 2009. "Production of Biodiesel Using a Continuous Gas–Liquid Reactor." *Bioresource technology* 100, no. 2, pp. 683–9. doi: 10.1016/j.biortech.2008.06.037.

Benjapornkulaphong, S., C. Ngamcharussrivichai, and K. Bunyakiat, 2009. "Al_2O_3 Supported Alkali and Alkali Earth Metal Oxides for Transesterification of Palm Kernel Oil and Coconut Oil." *Chemical Engineering Journal* 145, no. 3, pp. 468–74. doi: 10.1016/j.cej.2008.04.036.

Berchmans, H.J., and S. Hirata 2008. "Biodiesel Production from Crude Jatropha Curcas L. Seed Oil with a High Content of Free Fatty Acids." *Bioresource Technology* 99, no. 6, pp. 1716–21. doi: 10.1016/j.biortech.2007.03.051.

Bernardes, O.L., J.V. Bevilaqua, M.C. Leal, D.M. Freire, and M.A. Langone 2007. "Biodiesel Fuel Production by the Transesterification Reaction of Soybean Oil Using Immobilized Lipase." In *Applied Biochemistry and Biotechnology*, pp. 105–14. Humana Press.

Berrios, M., and R.L. Skelton. 2008. "Comparison of Purification Methods for Biodiesel." *Chemical Engineering Journal* 144, no. 3, pp. 459–65. doi: 10.1016/j.cej.2008.07.019.

Biofluidtech.com, http://www.biofluidtech.com/equipment.html

Birdwell, J.F., H.L. Jennings, J. Mcfarlane, and C. Tsouris. 2009. Integrated Reactor and Centrifugal Separator and Uses Thereof. US Patent 20,090,293,346. Washington, DC: U.S. Patent and Trademark Office.

Bo, X., X. Guomin, C. Lingfeng, W. Ruiping, and G. Lijing. 2007. "Transesterification of Palm Oil with Methanol to Biodiesel over a KF/Al2O3 Heterogeneous Base Catalyst." *Energy & Fuels* 21, no. 6, pp. 3109–12. doi: 10.1021/ef7005035.

Bojan, S.G., and S.K. Durairaj. 2012. "Producing Biodiesel from High Free Fatty Acid Jatropha Curcas Oil by a Two Step Method-An Indian Case Study." *Journal of Sustainable Energy and Environment* 3, no. 2, pp. 63–6.

Bolognini, M., F. Cavani, D. Scagliarini, C. Flego, C. Perego, and M. Saba. 2002. "Heterogeneous Basic Catalysts as Alternatives to Homogeneous Catalysts: Reactivity of Mg/Al Mixed Oxides in the Alkylation of m-Cresol with Methanol." *Catalysis Today* 75, no. 1, pp. 103–11. doi: 10.1016/S0920-5861(02)00050-0.

Bomb, C., K. McCormick, E. Deurwaarder, and T. Kåberger. 2007. "Biofuels for Transport in Europe: Lessons from Germany and the UK." *Energy Policy* 35, no. 4, pp. 2256–67. doi: 10.1016/j.enpol.2006.07.008.

Boocock, D.G., S.K. Konar, V. Mao, and H. Sidi. 1996. "Fast One-phase Oil-rich Processes for the Preparation of Vegetable Oil Methyl Esters." *Biomass and Bioenergy* 11, no. 1, pp. 43–50. doi: 10.1016/0961-9534(95)00111-5.

Boocock, D.G., S.K. Konar, V.Mao, C. Lee, and S. Buligan. 1998. "Fast Formation of High-purity Methyl Esters from Vegetable Oils." *Journal of the American Oil Chemists' Society* 75, no. 12, pp. 1167–72. doi: 10.1007/s11746-998-0307-1.

Bouaid, A., M. Martinez, and J. Aracil. 2007. "Long Storage Stability of Biodiesel from Vegetable and Used Frying Oils." *Fuel* 86, no. 16, pp. 2596–602. doi: 10.1016/j.fuel.2007.02.014.

Boz, N., N. Degirmenbasi, and D.M. Kalyon. 2009. "Conversion of Biomass to Fuel: Transesterification of Vegetable Oil to Biodiesel Using KF Loaded Nano-γ-Al$_2$O$_3$ as Catalyst." *Applied Catalysis B: Environmental* 89, no. 3, pp. 590–6. doi: 10.1016/j.apcatb.2009.01.026.

Bozbas, K. 2008. "Biodiesel as an Alternative Motor Fuel: Production and Policies in the European Union." *Renewable and Sustainable Energy Reviews* 12, no. 2, pp. 542–52. doi: 10.1016/j.rser.2005.06.001.

Bromberg, L., D.R. Cohn, and A. Rabinovich. 1999. *After treatment of Diesel Vehicle Emissions Using Compact Plasmatron Fuel Converter-catalyst Systems.* Cambridge, MA: Plasma Science and Fusion Center, Massachusetts Institute of Technology.

Bruwer, J.J., B. Van der Boshoff, F.J.C. Hugo, J. Fuls, C. Hawkins, A.N Van der Walt, and L.M. Du Plessis. 1981. "Utilization of Sunflower Seed Oil as a Renewable Fuel for Diesel Engines." ASAE National Energy Symposium, Vol. 4, ASAE Publ.

Bunce, M., D. Snyder, G. Adi, C. Hall, and G. Shaver. January 2010. "Optimization of the Performance and Emissions of Soy Biodiesel Blends in a Modern Diesel Engine." In *ASME 2010 Internal Combustion Engine Division Fall Technical Conference*, pp. 129–38. American Society of Mechanical Engineers.

Cai, L. 2014. "Thin Layer Chromatography." *Current Protocols Essential Laboratory Techniques* 6-3.

Campanati, M., G. Fornasari, and A. Vaccari. 2003. "Fundamentals in the Preparation of Heterogeneous Catalysts." *Catalysis Today* 77, no. 4, pp. 299–314. doi: 10.1016/S0920-5861(02)00375-9.

Canakci, M., and H. Sanli. 2008. "Biodiesel Production from Various Feedstocks and Their Effects on the Fuel Properties." *Journal of Industrial Microbiology & Biotechnology* 35, no. 5, pp. 431–41. doi: 10.1007/s10295-008-0337-6.

Canakci, M., and J. Van Gerpen. 1999. "Biodiesel Production Via Acid Catalysis." *Transactions of the American Society of Agricultural Engineers* 42, no. 5, pp. 1203–10. doi: 10.13031/2013.13285.

Canakci, M., and J. Van Gerpen. 2001. "Biodiesel Production from Oils and Fats with High Free Fatty Acids." *Transactions of the American Society of Agricultural Engineers* 44, no. 6, pp. 1429–36. doi: 10.13031/2013.7010.

Canakci, M., and J. Van Gerpen. 2003. "A Pilot Plant to Produce Biodiesel from High Free Fatty Acid Feedstocks." *Transactions of the American Society of Agricultural Engineers* 46, no. 4, pp. 945–54. doi: 10.13031/2013.13949.

Cao, P., M.A. Dubé, and A.Y. Tremblay. 2008. "High-purity Fatty Acid Methyl Ester Production from Canola, Soybean, Palm, and Yellow Grease Lipids by Means of a Membrane Reactor." *Biomass and Bioenergy* 32, no. 11, pp. 1028–36. doi: 10.1016/j.biombioe.2008.01.020.

Cao, W., H. Han, and J. Zhang. 2005. "Preparation of Biodiesel from Soybean Oil Using Supercritical Methanol and Co-solvent." *Fuel* 84, no. 4, pp. 347–51. doi: 10.1016/j.fuel.2004.10.001.

Carvalho, M.S., M.A. Mendonça, D.M. Pinho, I.S. Resck, and P.A. Suarez. 2012. "Chromatographic Analyses of Fatty Acid Methyl Esters by HPLC-UV and GC-FID." *Journal of the Brazilian Chemical Society* 23, no. 4, pp. 763–9. doi: 10.1590/S0103-50532012000400023.

Cavalcanti, E.H.D.S., M.B. Dantas, A.R. Albuquerque, L.E.B. Soledade, N. Queiroz, A.S. Maia, M.G. Santos, A.L. Barro, and A.G. Souza. 2011. "Biodiesel from Soybean Oil, Castor Oil and Their Blends." *Journal of Thermal Analysis and Calorimetry* 106, no. 2, pp. 607–11. doi: 10.1007/s10973-011-1410-3.

Centi, G., and S. Perathoner. 2003. "Novel Catalyst Design for Multiphase Reactions." *Catalysis Today* 79, pp. 3–13. doi: 10.1016/S0920-5861(03)00036-1.

Chan, E.S., B.B. Lee, P. Ravindra, and D. Poncelet. 2009. "Prediction Models for Shape and Size of Ca-alginate Macrobeads Produced Through Extrusion–Dripping Method." *Journal of Colloid and Interface Science* 338, no. 1, pp. 63–72. doi: 10.1016/j.jcis.2009.05.027.

Charles Ross & Son Company. http://www.compositesworld.com/suppliers/charles

Chattopadhyay, S., and R. Sen. 2013. "Fuel Properties, Engine Performance and Environmental Benefits of Biodiesel Produced by a Green Process." *Applied Energy* 105, pp. 319–26. doi: 10.1016/j.apenergy.2013.01.003.

Chen, K.J., and Y.S. Chen. 2014. "Intensified Production of Biodiesel Using a Spinning Disk Reactor." *Chemical Engineering and Processing: Process Intensification* 78, pp. 67–72. doi: 10.1016/j.cep.2014.02.009.

Chhetri, A.B., M.S. Tango, S.M. Budge, K.C. Watts, and M.R. Islam. 2008. "Non-edible Plant Oils as New Sources for Biodiesel Production." *International Journal of Molecular Sciences* 9, no. 2, pp. 169–80. doi: 10.3390/ijms9020169.

Chisti, Y. 2007. "Biodiesel from Microalgae." *Biotechnology Advances* 25, no. 3, pp. 294–306. doi: 10.1016/j.biotechadv.2007.02.001.

Choedkiatsakul, I., K. Ngaosuwan, and S. Assabumrungrat. 2013. "Application of Heterogeneous Catalysts for Transesterification of Refined Palm Oil in Ultrasound-assisted Reactor." *Fuel Processing Technology* 111, pp. 22–8. doi: 10.1016/j.fuproc.2013.01.015.

Choi, C.Y., G.R. Bower, and R.D. Reitz. 1997. Effects of Biodiesel Blended Fuels and Multiple Injections on DI Diesel Engines (No. 970218). SAE Technical Paper.

Chorkendorff, I., and J.W. Niemantsverdriet. 2006. *Concepts of Modern Catalysis and Kinetics*. John Wiley & Sons.

Cintas, P.,S. Mantegna, E.C. Gaudino, and G. Cravotto. 2010. "A New Pilot Flow Reactor for High-intensity Ultrasound Irradiation. Application to the Synthesis of Biodiesel." *Ultrasonics Sonochemistry* 17, no. 6, pp. 985–9. doi: 10.1016/j.ultsonch.2009.12.003.

Cosgrove, J.P., D.F. Church, and W.A. Pryor. 1987. "The Kinetics of the Autoxidation of Polyunsaturated Fatty acids." *Lipids* 22, no. 5, pp. 299–304. doi: 10.1007/BF02533996.

Costello, R. 2006. "Tiny Reactors Aim for Big Role." *Chemical Processing* 69, no. 12, pp. 14–19.

Darnoko, D., M. Cheryan, and E.G. Perkins. 2000. "Analysis of Vegetable Oil Transesterification Products by Gel Permeation Chromatography." *Journal of Liquid Chromatography & Related Technologies* 23, no. 15, pp. 2327–35. doi: 10.1081/JLC-100100491.

De Almeida, R.M., L.K. Noda, N.S. Gonçalves, S.M. Meneghetti, and M.R. Meneghetti 2008. "Transesterification Reaction of Vegetable Oils, Using Superacid Sulfated TiO_2 Catalysts." *Applied Catalysis A: General*, 347, no. 1, pp. 100–105. doi: 10.1016/j.apcata.2008.06.006.

De Filippis, P., C. Giavarini, M. Scarsella, and M. Sorrentino. 1995. "Transesterification Processes for Vegetable Oils: A Simple Control Method of Methyl Ester Content." *Journal of the American Oil Chemists' Society* 72, no. 11, pp. 1399–404. doi: 10.1007/BF02546218.

De Jong, R., and R. Suijker. 2012. "GC Analysis of Biodiesels: Compliance with International Standards Using a Single System." *American Laboratory* 44, no. 9, pp. 18–21.

Demirbas, A. 2002. "Biodiesel from Vegetable Oils via Transesterification in Supercritical Methanol." *Energy Conversion and Management* 43, no. 17, pp. 2349–56. doi: 10.1016/S0196-8904(01)00170-4.

Demirbas, A. 2003. "Biodiesel Fuels from Vegetable Oils via Catalytic and Non-catalytic Supercritical Alcohol Transesterifications and Other Methods: A Survey." *Energy Conversion and Management* 44, no. 13, pp. 2093–109. doi: 10.1016/S0196-8904(02)00234-0.

Demirbas, A. 2005. "Biodiesel Production from Vegetable Oils via Catalytic and Non-catalytic Supercritical Methanol Transesterification Methods." *Progress in Energy and Combustion Science* 31, no. 5, pp. 466–87. doi: 10.1016/j.pecs.2005.09.001.

Demirbas, A. 2006. "Biodiesel Production via Non-catalytic SCF Method and Biodiesel Fuel Characteristics." *Energy Conversion and Management* 47, no. 15, pp. 2271–82. doi: 10.1016/j.enconman.2005.11.019.

Demirbas, A. 2007a. "Biodiesel from Sunflower Oil in Supercritical Methanol with Calcium Oxide." *Energy Conversion and Management* 48, no. 3, pp. 937–41. doi: 10.1016/j.enconman.2006.08.004.

Demirbas, A. 2007b. "Importance of Biodiesel as Transportation Fuel." *Energy Policy* 35, no. 9, pp. 4661–70. doi: 10.1016/j.enpol.2007.04.003.

Demirbas, A. 2008a. *Biodiesel* Springer: London. pp. 111–9.

Demirbas, A. 2008b. "Biodiesel from Triglycerides via Transesterification." In *Biodiesel: A Realistic Fuel Alternative for Diesel Engines*. Springer: London. pp. 121–40.

Demirbas, A. 2009a. "Biodiesel from Waste Cooking Oil via Base-catalytic and Supercritical Methanol Transesterification." *Energy Conversion and Management* 50, no. 4, pp. 923-7. doi: 10.1016/j.enconman.2008.12.023.

Demirbas, A. 2009b. "Progress and Recent Trends in Biodiesel Fuels." *Energy Conversion and Management* 50, no. 1, pp. 14–34. doi: 10.1016/j.enconman.2008.09.001.

Demirbas, M.F. 2011. "Biofuels from Algae for Sustainable Development." *Applied Energy* 88, no. 10, pp. 3473–80. doi: 10.1016/j.apenergy.2011.01.059.

Deshpande, A., G. Anitescu, P.A. Rice, and L.L. Tavlarides. 2010. "Supercritical Biodiesel Production and Power Cogeneration: Technical and Economic Feasibilities." *Bioresource Technology* 101, no. 6, pp. 1834–43. doi: 10.1016/j.biortech.2009.10.034.

Di Felice, R., D. De Faveri, P. De Andreis, and P. Ottonello. 2008. "Component Distribution Between Light and Heavy Phases in Biodiesel Processes." *Industrial & Engineering Chemistry Research* 47, no. 20, pp. 7862–7. doi: 10.1021/ie800510w.

Di Serio, M., M. Cozzolino, R. Tesser, P. Patrono, F. Pinzari, B. Bonelli, and E. Santacesaria. 2007a. "Vanadyl Phosphate Catalysts in Biodiesel Production." *Applied Catalysis A: General* 320, pp. 1–7. doi: 10.1016/j.apcata.2006.11.025.

Di Serio, M., M. Ledda, M. Cozzolino, G. Minutillo, R. Tesser, and E. Santacesaria. 2006. "Transesterification of Soybean Oil to Biodiesel by Using Heterogeneous Basic Catalysts." *Industrial & Engineering Chemistry Research* 45, no. 9, pp. 3009–14. doi: 10.1021/ie051402o.

Di Serio, M., R. Tesser, M. Dimiccoli, F. Cammarota, M. Nastasi, and E. Santacesaria. 2005. "Synthesis of Biodiesel via Homogeneous Lewis Acid Catalyst." *Journal of Molecular Catalysis A: Chemical* 239, no. 1, pp. 111–5. doi: 10.1016/j.molcata.2005.05.041.

Di Serio, M., R. Tesser, L. Pengmei, and E. Santacesaria. 2007b. "Heterogeneous Catalysts for Biodiesel Production." *Energy & Fuels* 22, no. 1, pp. 207–17. doi: 10.1021/ef700250g.

Diehl, B., and G. Randel. 2007. "Analysis of Biodiesel, Diesel and Gasoline by NMR Spectroscopy–A Quick and Robust Alternative to NIR and GC." *Lipid Technology* 19, no. 11, pp. 258–60. doi: 10.1002/lite.200700087.

D'ippolito, S.A., J.C. Yori, M.E. Iturria, C.L. Pieck, and C.R. Vera. 2007. "Analysis of a Two-step, Noncatalytic, Supercritical Biodiesel Production Process with Heat Recovery." *Energy & Fuels* 21, no. 1, pp. 339–46. doi: 10.1021/ef060183w.

Donaldson, K., X.Y. Li, and W. MacNee. 1998. "Ultrafine (Nanometre) Particle Mediated Lung Injury." *Journal of Aerosol Science* 29, no. 5, pp. 553–60. doi: 10.1016/S0021-8502(97)00464-3.

Dossat, V., D. Combes, and A. Marty. 1999. "Continuous Enzymatic Transesterification of High Oleic Sunflower Oil in a Packed Bed Reactor: Influence of the Glycerol Production." *Enzyme and Microbial Technology* 25, no. 3, pp. 194–200. doi: 10.1016/S0141-0229(99)00026-5.

Dragone, G., B.D. Fernandes, A.A. Vicente, and J.A. Teixeira. 2010. "Third Generation Biofuels from Microalgae". In *Current research, technology and education topics in applied microbiology and microbial biotechnology.* Formatex

Dubé, M.A., A.Y. Tremblay, and J. Liu. 2007. "Biodiesel Production Using a Membrane Reactor." *Bioresource Technology* 98, no. 3, pp. 639–47. doi: 10.1016/j.biortech.2006.02.019.

Dunford, N.T. 2007. *Biodiesel Production Techniques.* Oklahoma Cooperative Extension Service, Division of Agricultural Sciences and Natural Resources.

Dunn, R.O. 2008. "Antioxidants for Improving Storage Stability of Biodiesel." *Biofuels, Bioproducts and Biorefining* 2, no. 4, pp. 304–18. doi: 10.1002/bbb.83.

Dunn, R.O., and M.O. Bagby. 1995. "Low-temperature Properties of Triglyceride-based Diesel Fuels: Transesterified Methyl Esters and Petroleum Middle Distillate/ester Blends." *Journal of the American Oil Chemists' Society* 72, no. 8, pp. 895–904. doi: 10.1007/BF02542067.

Dunn, R.O., G. Knothe, and M.O. Bagby. 1997. "Recent Advances in the Development of Alternative Diesel Fuel from Vegetable Oil and Animal Fats." *Recent Research Developments in Oil Chemistry* 1, pp. 31–56.

Dunn, R.O., M.W. Shockley, and M.O. Bagby. 1996. "Improving the Low-temperature Properties of Alternative Diesel Fuels: Vegetable Oil-derived Methyl Esters." *Journal of the American Oil Chemists' Society* 73, no. 12, pp. 1719–28. doi: 10.1007/BF02517978.

Edith, O., R.B. Janius, and R. Yunus. 2012. "Factors Affecting the Cold Flow Behavior of Biodiesel and Methods for Improvement–A Review." *Pertanika Journal of Science & Technology* 20, no. 1, pp. 1–14.

Eevera, T., K. Rajendran, and S. Saradha. 2009. "Biodiesel Production Process Optimization and Characterization to Assess the Suitability of the Product for Varied Environmental Conditions." *Renewable Energy* 34, no. 3, pp. 762–5. doi: 10.1016/j.renene.2008.04.006.

Ejigu, A., A. Asfaw, N. Asfaw, and P. Licence. 2010. "Moringa Stenopetala Seed Oil as a Potential Feedstock for Biodiesel Production in Ethiopia." *Green Chemistry* 12, no. 2, pp. 316–20. doi: 10.1039/B916500B.

El Sherbiny, S.A., A.A. Refaat, and S.T. El Sheltawy. 2010. "Production of Biodiesel Using the Microwave Technique." *Journal of Advanced Research* 1, no. 4, pp. 309–14. doi: 10.1016/j.jare.2010.07.003.

Elst, K., J. Sijben, and L. Van Ginneken. 2011. Method for Preparing Fatty Acid Esters with Alcohol Recycling. US Patent 8,076,498. Washington, DC: U.S. Patent and Trademark Office.

Encinar, J.M., J.F. Gonzalez, and A. Rodríguez-Reinares. 2005. "Biodiesel from Used Frying Oil. Variables Affecting the Yields and Characteristics of the Biodiesel." *Industrial & Engineering Chemistry Research* 44, no. 15, pp. 5491–9. doi: 10.1021/ie040214f.

eXtension; the American researched based learning network, *Used and Waste Oil and Grease for Biodiesel* - eXtension

Faccini, C.S., M.E.D. Cunha, M.S.A. Moraes, L.C. Krause, M.C. Manique, M.R.A. Rodrigues, and E.B. Caramão. 2011. "Dry Washing in Biodiesel Purification: A Comparative Study of Adsorbents." *Journal of the Brazilian Chemical Society* 22, no. 3, pp. 558–63. doi: 10.1590/S0103-50532011000300021.

Fattah, R.A., N.A. Mostafa, M.S. Mahmoud, and W. Abdelmoez. 2014. "Recovery of Oil and Free Fatty Acids from Spent Bleaching Earth Using Sub-critical Water Technology Supported with Kinetic and Thermodynamic Study." *Advances in Bioscience and Biotechnology* 5, no. 3, pp. 261–72. doi: 10.4236/abb.2014.53033.

Fazal, M.A., A.S.M.A. Haseeb, and H.H. Masjuki. 2011. "Biodiesel Feasibility Study: An Evaluation of Material Compatibility; Performance; Emission and Engine Durability." *Renewable and Sustainable Energy Reviews* 15, no. 2, pp. 1314–24. doi: 10.1016/j.rser.2010.10.004.

Fedosov, S.N., N.A. Fernandes, and M.Y. Firdaus. 2014. "Analysis of Oil–biodiesel Samples by High Performance Liquid Chromatography Using the Normal Phase Column of New Generation and the Evaporative Light Scattering Detector." *Journal of Chromatography A* 1326, pp. 56–62. doi: 10.1016/j.chroma.2013.12.043.

Ferdous, K., M.R. Uddin, M.R. Uddin, M.R. Khan, and M.A. Islam. 2013. "Preparation and Optimization of Biodiesel Production from Mixed Feedstock Oil." *Nature* 1, no. 4, pp. 62–66. doi: 10.12691/ces-1-4-3.

Ferrari, R.A., V.D.S. Oliveira, and A. Scabio. 2005. "Oxidative Stability of Biodiesel from Soybean Oil Fatty Acid Ethyl Esters." *Scientia Agricola* 62, no. 3, pp. 291–5. doi: 10.1590/S0103-90162005000300014.

Ferretti, C.A., R.N. Olcese, C.R. Apesteguía, and J.I. Di Cosimo. 2009. "Heterogeneously-Catalyzed Glycerolysis of Fatty Acid Methyl Esters: Reaction Parameter Optimization." *Industrial & Engineering Chemistry Research* 48, no. 23, pp. 10387–94.

Fillières, R., B. Benjelloun-Mlayah, and M. Delmas. 1995. "Ethanolysis of Rapeseed Oil: Quantitation of Ethyl Esters, Mono-, Di-, and Triglycerides and Glycerol by High-performance Size-exclusion Chromatography." *Journal of the American Oil Chemists' Society* 72, no. 4, pp. 427–32. doi: 10.1007/BF02636083.

Fjerbaek, L., K.V. Christensen, and B. Norddahl. 2009. "A Review of the Current State of Biodiesel Production Using Enzymatic Transesterification." *Biotechnology and Bioengineering* 102, no. 5, pp. 1298–315. doi: 10.1002/bit.22256.

Fontana, J.D., G. Zagonel, W.W. Vechiatto, B.J. Costa, J.C. Laurindo, R. Fontana, and F.M. Lanças. 2009. "Simple TLC-Screening of Acylglycerol Levels in Biodiesel as an Alternative to GC Determination." *Journal of Chromatographic Science* 47, no. 9, pp. 844–6. doi: 10.1093/chromsci/47.9.844.

Freedman, B.E.H.P., E.H. Pryde, and T.L. Mounts. 1984. "Variables Affecting the Yields of Fatty Esters from Transesterified Vegetable Oils." *Journal of the American Oil Chemists Society* 61, no. 10, pp. 1638–43. doi: 10.1007/BF02541649.

Freedman, B., R.O. Butterfield, and E.H. Pryde. 1986. "Transesterification Kinetics of Soybean Oil 1." *Journal of the American Oil Chemists' Society* 63, no. 10, pp. 1375–80. doi: 10.1007/BF02679606.

Freedman, B., W.F. Kwolek, and Pryde. E.H. 1986. "Quantitation in the Analysis of Transesterified Soybean Oil by Capillary Gas Chromatography 1." *Journal of the American Oil Chemists' Society* 63, no. 10, pp. 1370–5. doi: 10.1007/BF02679605.

Freese, U., F. Heinrich, and F. Roessner. 1999. "Acylation of Aromatic Compounds on H-Beta Zeolites." *Catalysis Today* 49, no. 1, pp. 237–44. doi: 10.1016/S0920-5861(98)00429-5.

Fukuda, H., A. Kondo, and H. Noda. 2001. "Biodiesel Fuel Production by Transesterification of Oils." *Journal of Bioscience and Bioengineering* 92, no. 5, pp. 405–16. doi: 10.1016/S1389-1723(01)80288-7.

Furuta, S., H. Matsuhashi, and K. Arata. 2004. "Biodiesel Fuel Production with Solid Superacid Catalysis in Fixed Bed Reactor Under Atmospheric Pressure." *Catalysis Communications* 5, no. 12, pp. 721–3. doi: 10.1016/j.catcom.2004.09.001.

Furuta, S., H. Matsuhashi, and K. Arata. 2006. "Biodiesel Fuel Production with Solid Amorphous-Zirconia Catalysis in Fixed Bed Reactor." *Biomass and Bioenergy* 30, no. 10, pp. 870–3. doi: 10.1016/j.biombioe.2005.10.010.

Gadonneix, P., F.B. de Castro, N.F. de Medeiros, R. Drouin, C.P. Jain, Y.D. Kim, and C. Frei. 2010. Biofuels: Policies, Standards and Technologies. World Energy Council.

Gaita, R. 2006. "A Reversed Phase HPLC Method Using Evaporative Light Scattering Detection (ELSD) for Monitoring the Reaction and Quality of Biodiesel Fuels." *Lc Gc North America* 24, no. 9, pp. 51.

Ganesan, D., A. Rajendran, and V. Thangavelu. 2009. "An Overview on the Recent Advances in the Transesterification of Vegetable Oils for Biodiesel Production Using Chemical and Biocatalysts." *Reviews in Environmental Science and Bio/technology* 8, no. 4, pp. 367–94. doi: 10.1007/s11157-009-9176-9.

Garba, M.U., M. Alhassan, and S. Abdulsalami. 2006. "A Review of Advances and Quality Assessment of Biofuels." *Leonardo Journal of Sciences*, no. 9, pp. 167–78.

Garcez, C.A.G., and J.N.D.S. Vianna. 2009. "Brazilian Biodiesel Policy: Social and Environmental Considerations of Sustainability." *Energy* 34, no. 5, pp. 645–54. doi: 10.1016/j.energy.2008.11.005.

Gelbard, G., O. Bres, R.M. Vargas, F. Vielfaure, and U.F. Schuchardt. 1995. "1 H Nuclear Magnetic Resonance Determination of the Yield of the Transesterification of Rapeseed Oil with Methanol." *Journal of the American Oil Chemists' Society* 72, no. 10, pp. 1239–41. doi: 10.1007/BF02540998.

Gerpen, J.V. 2005. "Biodiesel Processing and Production." *Fuel Processing Technology* 86, no. 10, pp. 1097–107.doi: 10.1016/j.fuproc.2004.11.005.

Gerpen, V.J., G. Knothe. 2005. "Basics of the Transesterification Reaction." In *The Biodiesel Handbook*, eds. G. Knothe, J.V. Gerpen, and J. Krahl, pp. 26–41. Urbana, IL: AOCS Press.

Ghanem, A. 2003. "The Utility of Cyclodextrins in Lipase-Catalyzed Transesterification in Organic Solvents: Enhanced Reaction Rate and Enantioselectivity." *Organic & Biomolecular Chemistry* 1, no. 8, pp. 1282–91. doi: 10.1039/B301086D.

Giwa, S., L.C. Abdullah, and N.M. Adam. 2010. "Investigating "Egusi"(Citrullus Colocynthis L.) Seed Oil as Potential Biodiesel Feedstock." *Energies* 3, no. 4, pp. 607–18. doi:10.3390/en3040607.

Godfray, H.C.J., J.R. Beddington, I.R. Crute, L. Haddad, D. Lawrence, J.F. Muir, and C. Toulmin. 2010. "Food Security: The Challenge of Feeding 9 Billion People." *Science* 327, no. 5967, pp. 812–18. doi: 10.1126/science.1185383.

Goldemberg, J., and P. Guardabassi. 2009. "Are Biofuels a Feasible Option?" *Energy Policy* 37, no. 1, pp. 10–14. doi: 10.1016/j.enpol.2008.08.031.

Gomes, M.C.S., P.A. Arroyo, and N.C. Pereira. 2011. "Biodiesel Production from Degummed Soybean Oil and Glycerol Removal Using Ceramic Membrane." *Journal of Membrane Science* 378, no. 1, pp. 453–61. doi: 10.1016/j.memsci.2011.05.033.

Gomes, M.C.S., N.C. Pereira, and S.T.D.D. Barros. 2010. "Separation of Biodiesel and Glycerol Using Ceramic Membranes." *Journal of Membrane Science* 352, no. 1, pp. 271–6. doi: 10.1016/j.memsci.2010.02.030.

Gordon, R., I. Gorodnitsky, and V. Grichko. 2014. Process for Producing Biodiesel Through Lower Molecular Weight Alcohol-Targeted Cavitation. US Patent 20,140,059,922. Washington, DC: U.S. Patent and Trademark Office.

Gude, V.G., P. Patil, E. Martinez-Guerra, S. Deng, and N. Nirmalakhandan. 2013. "Microwave Energy Potential for Biodiesel Production." *Sustainable Chemical Processes* 1, no. 1, pp. 5. doi: 10.1186/2043-7129-1-5.

Gutierrez-Ortiz, J.I., R. López-Fonseca, C.G.O. de Elguea, M.P. González-Marcos, and J.R. González-Velasco. 2000. "Mass Transfer Studies in the Hydrogenation of Methyl Oleate over a Ni/SiO2 Catalyst in the Liquid Phase." *Reaction Kinetics and Catalysis Letters* 70, no. 2, pp. 341–8. doi: 10.1023/A:1010357420353.

Haas, M.J., A.J. McAloon, W.C. Yee, and T.A. Foglia. 2006. "A Process Model to Estimate Biodiesel Production Costs." *Bioresource Technology* 97, no. 4, pp. 671–8. doi: 10.1016/j.biortech.2005.03.039.

Hada, S., C.C. Solvason, and M.R. Eden. 2014. "Characterization-Based Molecular Design of Biofuel Additives Using Chemometric and Property Clustering Techniques." *Process and Energy Systems Engineering* 2, Article 20. doi: 10.3389/fenrg.2014.00020.

Hanh, H.D., N.T. Dong, K. Okitsu, R. Nishimura, and Y. Maeda. 2009a. "Biodiesel Production by Esterification of Oleic Acid with Short-chain Alcohols Under Ultrasonic Irradiation Condition." *Renewable Energy* 34, no. 3, pp. 780–783. doi: 10.1016/j.renene.2008.04.001.

Hanh, H.D., N.T. Dong, K. Okitsu, R. Nishimura, and Y. Maeda. 2009b. "Biodiesel Production Through Transesterification of Triolein with Various Alcohols in an Ultrasonic Field." *Renewable Energy* 34, no. 3, pp. 766–8. doi: 10.1016/j.renene.2008.04.007.

Harvey, A.P., M.R. Mackley, and T. Seliger. 2003. "Process Intensification of Biodiesel Production Using a Continuous Oscillatory Flow Reactor." *Journal of Chemical Technology and Biotechnology* 78, no. (2–3), pp. 338–41. doi: 10.1002/jctb.782.

Hatti-Kaul, R. 2010. "Downstream Processing in Industrial Biotechnology." In *Industrial Biotechnology: Sustainable Growth and Economic Success*, eds. W Soetaert, and E.J. Vandamme, pp. 279–321.

Hayyan, M., F.S. Mjalli, M.A. Hashim, and I.M. AlNashef. 2010. "A Novel Technique for Separating Glycerine from Palm Oil-based Biodiesel Using Ionic Liquids." *Fuel Processing Technology* 91, no. 1, pp. 116–20. doi: 10.1016/j.fuproc.2009.09.002.

He, B.B., A.P. Singh, and J.C. Thompson. 2005. "Experimental Optimization of a Continuous-flow Reactive Distillation Reactor for Biodiesel Production." *Transactions of the American Society of Agricultural Engineers* 48, no. 6, pp. 2237–43. doi: 10.13031/2013.20071.

He, B.B., A.P. Singh, and J.C. Thompson. 2007. "Function and Performance of a Pre-reactor to a Reactive Distillation Column for Biodiesel Production." *Transactions of the ASABE* 50, no. 1, pp. 123–8. doi: 10.13031/2013.22383.

He, B., A.P. Singh, and J.C. Thompson. 2006. "A Novel Continuous-flow Reactor Using Reactive Distillation Technique for Biodiesel Production." *Transactions of the ASABE* 49, no. 1, pp. 107–12. doi: 10.13031/2013.20218.

Heimann, D. 2003. "Process Intensification through the Combined Use of Process Simulation and Miniplant Technology." *Computer Aided Chemical Engineering* 14, pp. 155–60. doi: 10.1016/S1570-7946(03)80107-4.

Hill, J., E. Nelson, D. Tilman, S. Polasky, and D. Tiffany. 2006. "Environmental, Economic, and Energetic Costs and Benefits of Biodiesel and Ethanol Biofuels." *Proceedings of the National Academy of Sciences* 103, no. 30, pp. 11206–10. doi: 10.1073/pnas.0604600103.

Hoekman, S.K., A. Broch, C. Robbins, E. Ceniceros, and M. Natarajan. 2012. "Review of Biodiesel Composition, Properties, and Specifications." *Renewable and Sustainable Energy Reviews* 16, no. 1, pp. 143–69. doi: 10.1016/j.rser.2011.07.143.

Holman, R.T., and P.R. Edmondson. 1956. "Near-infrared Spectra of Fatty Acids and Some Related Substances." *Analytical chemistry* 28, no. 10, pp. 1533–8. doi: 10.1021/ac60118a010.

Howell, S. 1997. US Biodiesel Standards-An Update of Current Activities (No. 971687). SAE Technical Paper.

Hunt, S. 2007. "Potential Challenges and Risks of Bioenergy Production for Developing Countries." *Agriculture & Rural Development* 14, no. 2, pp. 30–33.

Hydro Dynamics, http://www.globalspec.com/supplier/profile/HydroDynamics

Islam, A., Y.H. Taufiq-Yap, C.M. Chu, E.S. Chan, and P. Ravindra. 2013a. "Studies on Design of Heterogeneous Catalysts for Biodiesel Production." *Process Safety and Environmental Protection* 91, no. 1, pp. 131–44. doi: 10.1016/j.psep.2012.01.002.

Islam, A., Y.H. Taufiq-Yap, C.M. Chu, P. Ravindra, and E.S. Chan. 2013b. "Transesterification of Palm Oil Using KF and Nano3 Catalysts Supported on Spherical Millimetric γ-Al2O3." *Renewable Energy* 59, pp. 23–29. doi: 10.1016/j.renene.2013.01.051.

Iso, M., B. Chen, M. Eguchi, T. Kudo, and S. Shrestha. 2001. "Production of Biodiesel Fuel from Triglycerides and Alcohol Using Immobilized Lipase." *Journal of Molecular Catalysis B: Enzymatic* 16, no. 1, pp. 53–58. doi: 10.1016/S1381-1177(01)00045-5.

Jarrah, N.A., J.G. van Ommen, and L. Lefferts. 2004. "Immobilization of Carbon Nanofibers (cnfs); A New Structured Catalyst Support." *Preprints of Papers-American Chemical Society, Division of Fuel Chemistry* 49, no. 2, pp. 881.

Jaruwat, P., S. Kongjao, and M. Hunsom. 2010. "Management of Biodiesel Wastewater by the Combined Processes of Chemical Recovery and Electrochemical Treatment." *Energy Conversion and Management* 51, no. 3, pp. 531–7. doi: 10.1016/j.enconman.2009.10.018.

Jayasinghe, T.K., P. Sungwornpatansakul, and K. Yoshikawa. 2014. "Enhancement of Pretreatment Process for Biodiesel Production from Jatropha Oil Having High Content of Free Fatty Acids." *International Journal of Energy Engineering* 4, no. 3, pp. 118–26.

Jennings, M.J. 2007. Bioenergy: Fueling the Future. *Drake J. Agric. L.*, 12, 205.

Ji, J., J. Wang, Y. Li, Y. Yu, and Z. Xu. 2006. "Preparation of Biodiesel with the Help of Ultrasonic and Hydrodynamic Cavitation." *Ultrasonics* 44, pp. e411–e414. doi: 10.1016/j.ultras.2006.05.020.

Jiang, L.Y., H. Chen, Y.C. Jean, and T.S. Chung. 2009. "Ultrathin Polymeric Inter-penetration Network with Separation Performance Approaching Ceramic Membranes for Biofuel." *AIChE Journal* 55, no. 1, pp. 75–86. doi: 10.1002/aic.11652.

Jin, F., K. Kawasaki, H. Kishida, K. Tohji, T. Moriya, and H. Enomoto. 2007. "NMR Spectroscopic Study on Methanolysis Reaction of Vegetable Oil." *Fuel* 86, no. 7, pp. 1201–7. doi: 10.1016/j.fuel.2006.10.013.

Joshi, H.C., J. Toler, and T. Walker. 2008. "Optimization of Cottonseed Oil Ethano-lysis to Produce Biodiesel High in Gossypol Content." *Journal of the American Oil Chemists' Society* 85, no. 4, pp. 357–63. doi: 10.1007/s11746-008-1200-7.

Kapilan, N. 2012. "Production of Biodiesel from Vegetable Oil Using Microware Irradiation." *Acta Polytechnica* 52, no. 1.

Karmakar, A., S. Karmakar, and S. Mukherjee. 2010. "Properties of Various Plants and Animals Feedstocks for Biodiesel Production." *Bioresource Technology* 101, no. 19, pp. 7201–10. doi: 10.1016/j.biortech.2010.04.079.

Kathrin Hielscher, Hielscher Ultrasonics GmbH, Germany, http://www.hielscher.com/biodiesel_transesterification_01.htm

Kawashima, A., Matsubara, K., and Honda, K. 2008. "Development of Hetero-geneous Base Catalysts for Biodiesel Production." *Bioresource Technology* 99(9), 3439–3443.

Keil, F. (Ed.). 2007. *Modeling of Process Intensification*. New York: John Wiley & Sons.

KianHee, K., S.M. Yasir, and K. Kudumpor. 2012. "Development of Fast Check Test Kit for Biodiesel Quality Monitoring." *International Journal of Chem-ical Engineering and Applications* 3, no. 5, pp. 307–10. doi: 10.7763/IJCEA.2012.V3.206.

Kim, D., J. Choi, G.J. Kim, S.K. Seol, Y.C. Ha, M. Vijayan, and S.S. Park. 2011. "Microwave-accelerated Energy-efficient Esterification of Free Fatty Acid with a Heterogeneous Catalyst." *Bioresource Technology* 102, no. 3, pp. 3639–41. doi: 10.1016/j.biortech.2010.11.067.

Kim, H.J., B.S. Kang, M.J. Kim, Y.M. Park, D.K. Kim, J.S. Lee, and K.Y. Lee. 2004. "Transesterification of Vegetable Oil to Biodiesel Using Heteroge-neous Base Catalyst." *Catalysis Today* 93, pp. 315–20. doi: 10.1016/j.cat-tod.2004.06.007.

Knothe, G. 1999. "Rapid Monitoring of Transesterification and Assessing Bio-diesel Fuel Quality by Near-infrared Spectroscopy Using a Fiber-optic Probe." *Journal of the American Oil Chemists' Society* 76, no. 7, pp. 795–800. doi: 10.1007/s11746-999-0068-5.

Knothe, G. 2000. "Monitoring a Progressing Transesterification Reaction by Fiber-optic Near Infrared Spectroscopy with Correlation to 1H Nuclear Mag-netic Resonance Spectroscopy." *Journal of the American Oil Chemists' Soci-ety* 77, no. 5, pp. 489–93. doi: 10.1007/s11746-000-0078-5.

Knothe, G. 2001. "Analytical Methods Used in the Production and Fuel Quality Assessment of Biodiesel." *Transactions of the ASAE* 44, no. 2, pp. 193–200. doi: 10.13031/2013.4740.

Knothe, G. 2005. "Dependence of Biodiesel Fuel Properties on the Structure of Fatty Acid Alkyl Esters." *Fuel Processing Technology* 86, no. 10, pp. 1059–70.

Knothe, G. 2006. "Analysis of Oxidized Biodiesel by 1H-NMR and Effect of Contact Area with Air." *European Journal of Lipid Science and Technology* 108, no. 6, pp. 493–500. doi: 10.1002/ejlt.200500345.

Knothe, G. 2007. "Some Aspects of Biodiesel Oxidative Stability." *Fuel Processing Technology* 88, no. 7, pp. 669–77. doi: 10.1016/j.fuproc.2007.01.005.

Knothe, G. 2008. "'Designer' Biodiesel: Optimizing Fatty Ester Composition to Improve Fuel Properties." *Energy & Fuels* 22, no. 2, pp. 1358–64. doi: 10.1021/ef700639e.

Knothe, G. 2010. "Biodiesel and Renewable Diesel: A Comparison." *Progress in Energy and Combustion Science* 36, no. 3, pp. 364–73. doi: 10.1016/j.pecs.2009.11.004.

Knothe, G., and J.A. Kenar. 2004. "Determination of the Fatty Acid Profile by 1H–NMR Spectroscopy." *European Journal of Lipid Science and Technology* 106, no. 2, pp. 88–96. doi: 10.1002/ejlt.200300880.

Knothe, G., J. Van Gerpen, and J. Krahl. 2005. *The biodiesel handbook* 1. Champaign, IL: AOCS press.

Knothe, G., and K.R. Steidley. 2005. "Kinematic Viscosity of Biodiesel Fuel Components and Related Compounds. Influence of Compound Structure and Comparison to Petrodiesel Fuel Components." *Fuel* 84, no. 9, pp. 1059–65. doi: 10.1016/j.fuel.2005.01.016.

Knothe, G., and K.R. Steidley. 2007. "Kinematic Viscosity of Biodiesel Components (Fatty Acid Alkyl Esters) and Related Compounds at Low Temperatures." *Fuel* 86, no. 16, pp. 2560–7. doi: 10.1016/j.fuel.2007.02.006.

Knothe, G., R.O. Dunn, and M.O. Bagby. January 1997. "Biodiesel: The Use of Vegetable Oils and Their Derivatives as Alternative Diesel Fuels." *ACS symposium series*, 666, pp. 172–208. Washington, DC: American Chemical Society.

Kocherginsky, N.M., Q. Yang, and L. Seelam. 2007. "Recent Advances in Supported Liquid Membrane Technology." *Separation and Purification Technology* 53, no. 2, pp. 171–7. doi: 10.1016/j.seppur.2006.06.022.

Koh, L.P., and J. Ghazoul. 2008. "Biofuels, Biodiversity, and People: Understanding the Conflicts and Finding Opportunities." *Biological Conservation* 141, no. 10, pp. 2450–60. doi: 10.1016/j.biocon.2008.08.005.

Körbitz, W. 1999. "Biodiesel Production in Europe and North America, an Encouraging Prospect." *Renewable Energy* 16, no. 1, pp. 1078–83.

Krahl, J., J. Bünger, O. Schröder, A. Munack, and G. Knothe. 2002. "Exhaust Emissions and Health Effects of Particulate Matter from Agricultural Tractors Operating on Rapeseed Oil Methyl Ester." *Journal of the American Oil Chemists' Society* 79, no. 7, pp. 717–24. doi: 10.1007/s11746-002-0548-9.

Krisnangkura, K., and R. Simamaharnnop. 1992. "Continuous Transmethylation of Palm Oil in an Organic Solvent." *Journal of the American Oil Chemists' Society* 69, no. 2, pp. 166–9. doi: 10.1007/BF02540569.

Kuhn, T., and M. Pickhardt. 2009. "Biofuels, Innovations, and Endogenous Growth." In *New Developments in Schumpetarian Economics*, eds. U. Canter, A. Greiner, T, Kuhn, Pyka, pp. 182–96. Cheltenham: Edward Elgar

Kumar, D., G. Kumar, and C.P. Singh. 2010. "Fast, Easy Ethanolysis of Coconut Oil for Biodiesel Production Assisted by Ultrasonication." *Ultrasonics Sonochemistry* 17, no. 3, pp. 555–9. doi: 10.1016/j.ultsonch.2009.10.018.

Kumar, R., G. Ravi Kumar, and N. Chandrashekar. 2011. "Microwave Assisted Alkali-catalyzed Transesterification of Pongamia Pinnata Seed Oil for Biodiesel Production." *Bioresource Technology* 102, no. 11, pp. 6617–20. doi: 10.1016/j.biortech.2011.03.024.

Kumar, S., A. Chaube, and S.K. Jain. 2010. "Issues Pertaining to Substitution of Diesel by Jatropha Biodiesel in India." *Journal of Environmental Research and Development* 4, no. 3, pp. 876–84.

Kumari, A., P. Mahapatra, V.K. Garlapati, and R. Banerjee. 2009. "Enzymatic Transesterification of Jatropha Oil." *Biotechnology for Biofuels* 2, no. 1, pp. 1–6. doi: 10.1186/1754-6834-2-1.

Kusdiana, D., and S. Saka. 2004. "Effects of Water on Biodiesel Fuel Production by Supercritical Methanol Treatment." *Bioresource Technology* 91, no. 3, pp. 289–95. doi: 10.1016/S0960-8524(03)00201-3.

Kusdiana, D., and S. Saka. 2001. "Kinetics of Transesterification in Rapeseed Oil to Biodiesel Fuel as Treated in Supercritical Methanol." *Fuel* 80, no. 5, pp. 693–8. doi: 10.1016/S0016-2361(00)00140-X.

Kwanchareon, P., A. Luengnaruemitchai, and S. Jai-In. 2007. "Solubility of a Diesel–Biodiesel–Ethanol Blend, its Fuel Properties, and its Emission Characteristics from Diesel Engine." *Fuel* 86, no. 7, pp. 1053–61. doi: 10.1016/j.fuel.2006.09.034.

Lam, M.K., and K.T. Lee. 2010. "Accelerating Transesterification Reaction with Biodiesel as Co-solvent: A Case Study for Solid Acid Sulfated Tin Oxide Catalyst." *Fuel* 89, no. 12, pp. 3866–70. doi: 10.1016/j.fuel.2010.07.005.

Lam, M.K., K.T. Lee, and A.R. Mohamed. 2010. "Homogeneous, Heterogeneous and Enzymatic Catalysis for Transesterification of High Free Fatty Acid Oil (Waste Cooking Oil) to Biodiesel: A Review." *Biotechnology Advances* 28, no. 4, pp. 500–18. doi: 10.1016/j.biotechadv.2010.03.002.

Lam, M.K., K.T. Tan, K.T. Lee, and A.R. Mohamed. 2009. "Malaysian Palm Oil: Surviving the Food Versus Fuel Dispute for a Sustainable Future." *Renewable and Sustainable Energy Reviews* 13, no. 6, pp. 1456–64. doi: 10.1016/j.rser.2008.09.009.

Leadbeater, N.E., and L.M. Stencel. 2006. "Fast, Easy Preparation of Biodiesel Using Microwave Heating." *Energy & Fuels* 20, no. 5, pp. 2281–83. doi: 10.1021/ef060163u.

Leite, J.G.D.B., J. Bijman, K. Giller, and M. Slingerland. 2013. "Biodiesel Policy for Family Farms in Brazil: One-size-fits-all?" *Environmental Science & Policy* 27, pp. 195–205. doi: 10.1016/j.envsci.2013.01.004.

Lengyel, J., Z. Cvengrošová, and J. Cvengroš. 2009. "Transesterification of Triacylglycerols over Calcium Oxide as Heterogeneous Catalyst." *Petroleum & coal* 51, no. 3, pp. 216–24.

Lertsathapornsuk, V., R. Pairintra, K. Aryusuk, and K. Krisnangkura. 2008. "Microwave Assisted in Continuous Biodiesel Production from Waste Frying Palm Oil and Its Performance in a 100 kW Diesel Generator." *Fuel Processing Technology* 89, no. 12, pp. 1330–6. doi: 10.1016/j.fuproc.2008.05.024.

Leung, D.Y.C., and Y. Guo. 2006. "Transesterification of Neat and Used Frying Oil: Optimization for Biodiesel Production." *Fuel Processing Technology* 87, no. 10, pp. 883–90. doi: 10.1016/j.fuproc.2006.06.003.

LeViness, S., A.L. Tonkovich, K. Jarosch, S. Fitzgerald, B. Yang, and J. McDaniel. 2011. *Improved Fischer-Tropsch Economics Enabled by Microchannel Technologyi.* Velocys.

Li, Y., M. Horsman, N. Wu, C.Q. Lan, and N. Dubois–Calero. 2008. "Biofuels from Microalgae." *Biotechnology Progress* 24, no. 4, pp. 815–20. doi: 10.1021/bp070371k.

Linko, Y.Y., M. Lämsä, X. Wu, E. Uosukainen, J. Seppälä, and P. Linko. 1998. "Biodegradable Products by Lipase Biocatalysis." *Journal of Biotechnology* 66, no. 1, pp. 41–50. doi: 10.1016/S0168-1656(98)00155-2.

Liu, K.S. 1994. "Preparation of Fatty Acid Methyl Esters for Gas-chromatographic Analysis of Lipids in Biological Materials." *Journal of the American Oil Chemists' Society* 71, no. 11, pp. 1179–87. doi: 10.1007/BF02540534.

Lotero, E., Y. Liu, D.E. Lopez, K. Suwannakarn, D.A. Bruce, and J.G. Goodwin. 2005. "Synthesis of Biodiesel via Acid Catalysis." *Industrial & Engineering Chemistry Research* 44, no. 14, pp. 5353–63. doi: 10.1021/ie049157g.

Lou, W.Y., M.H. Zong, and Z.Q. Duan. 2008. "Efficient Production of Biodiesel from High Free Fatty Acid-containing Waste Oils Using Various Carbohydrate-derived Solid Acid Catalysts." *Bioresource Technology* 99, no. 18, pp. 8752–8. doi: 10.1016/j.biortech.2008.04.038.

Loupy, A., A. Petit, M. Ramdani, C. Yvanaeff, M. Majdoub, B. Labiad, and D. Villemin. 1993. "The Synthesis of Esters Under Microwave Irradiation Using Dry-media Conditions." *Canadian Journal of Chemistry* 71, no. 1, pp. 90–95. doi: 10.1139/v93-013.

Lozano, P., N. Chirat, J. Graille, and D. Pioch. 1996. "Measurement of Free Glycerol in Biofuels." *Fresenius' Journal of Analytical Chemistry* 354, no. 3, pp. 319–22. doi: 10.1007/s0021663540319.

Lü, J., C. Sheahan, and P. Fu. 2011. "Metabolic Engineering of Algae for Fourth Generation Biofuels Production." *Energy & Environmental Science* 4, no. 7, pp. 2451–66. doi: 10.1039/C0EE00593B.

Ma, F., and M.A. Hanna. 1999. "Biodiesel Production: A Review." *Bioresource Technology* 70, no. 1, pp. 1–15. doi: 10.1016/S0960-8524(99)00025-5.

Ma, F., L.D. Clements, and M. Hanna. 1998. "The Effects of Catalyst, Free Fatty Acids, and Water on Transecterification of Beef Tallow." *Transactions of the ASABE* 41, no. 5, pp. 1261–4. doi: 10.13031/2013.17292.

Ma, F., L.D. Clements, and M.A. Hanna. 1999. "The Effect of Mixing on Transesterification of Beef Tallow." *Bioresource Technology* 69, no. 3, pp. 289–93. doi: 10.1016/S0960-8524(98)00184-9.

Ma, H., S. Li, B. Wang, R. Wang, and S. Tian. 2008. "Transesterification of Rape-seed Oil for Synthesizing Biodiesel by K/KOH/γ-Al2O3 as Heterogeneous Base Catalyst." *Journal of the American Oil Chemists' Society* 85, no. 3, pp. 263–70. doi: 10.1007/s11746-007-1188-4.

Mabaso, E.I., E. Van Steen, and M. Claeys. 2006. "Fischer-Tropsch Synthesis on Supported Iron Crystallites of Different Size." *Synthesis Gas Chemistry Conference*, Vol. 4, pp. 93–100. Hamburg, Germany: DGMK Tagungsbericht.

Maceiras, R., M. Vega, C. Costa, P. Ramos, and M.C. Márquez. 2009. "Effect of Methanol Content on Enzymatic Production of Biodiesel from Waste Frying Oil." *Fuel* 88, no. 11, pp. 2130–4. doi: 10.1016/j.fuel.2009.05.007.

Mackley, M.R. 1991. "Process Innovation Using Oscillatory Flow Within Baffled Tubes." *Chemical Engineering Research & Design* 69, no. A3, pp. 197–9.

MacLeod, C.S., A.P. Harvey, A.F. Lee, and K. Wilson. 2008. "Evaluation of the Activity and Stability of Alkali-doped Metal Oxide Catalysts for Application to an Intensified Method of Biodiesel Production." *Chemical Engineering Journal* 135, no. 1, pp. 63–70. doi: 10.1016/j.cej.2007.04.014.

Madras, G., C. Kolluru, and R. Kumar. 2004. "Synthesis of Biodiesel in Supercritical Fluids." *Fuel* 83, no. 14, pp. 2029–33. doi: 10.1016/j.fuel.2004.03.014.

Marchetti, J.M., V.U. Miguel, and A.F. Errazu. 2008. "Techno-economic Study of Different Alternatives for Biodiesel Production." *Fuel Processing Technology* 89, no. 8, pp. 740–48. doi: 10.1016/j.fuproc.2008.01.007.

Margaroni, D. 1998. "Fuel Lubricity." *Industrial Lubrication and Tribology* 50, no. 3, pp. 108–18. doi: 10.1108/00368799810218026.

Matsuhashi, H., H. Miyazaki, Y. Kawamura, H. Nakamura, and K. Arata. 2001. "Preparation of a Solid Superacid of Sulfated Tin Oxide with Acidity Higher than that of Sulfated Zirconia and its Applications to Aldol Condensation and Benzoylation1." *Chemistry of Materials* 13, no. 9, pp. 3038–42. doi: 10.1021/cm0104553.

Mavournin, K.H., D.H. Blakey, M.C. Cimino, M.F. Salamone, and J.A. Heddle. 1990. "The In Vivo Micronucleus Assay in Mammalian Bone Marrow and Peripheral Blood. A Report of the US Environmental Protection Agency Gene-Tox Program." *Mutation Research/Reviews in Genetic Toxicology* 239, no. 1, pp. 29–80. doi: 10.1016/0165-1110(90)90030-F.

McCarty, G.S., and P.S. Weiss. 1999. "Scanning Probe Studies of Single Nanostructures." *Chemical Reviews* 99, no. 7, pp. 1983–90. doi: 10.1021/cr970110x.

McCormick, R.L., M. Ratcliff, L. Moens, and R. Lawrence. 2007. "Several Factors Affecting the Stability of Biodiesel in Standard Accelerated Tests." *Fuel Processing Technology* 88, no. 7, pp. 651–7. doi: 10.1016/j.fuproc.2007.01.006.

McMichael, P. 2012. "Biofuels and the Financialization of the Global Food System." In *Food Systems Failure: The Global Food Crisis and the Future of Agriculture*, eds. C. Rosin, P. Stock, and H. Campbell, pp. 60–82. Abingdon, England: Earthscan.

McNeff, C.V., L.C. McNeff, B. Yan, D.T. Nowlan, M. Rasmussen, A.E. Gyberg, and T.R. Hoye. 2008. "A Continuous Catalytic System for Biodiesel

Production." *Applied Catalysis A: General* 343, no. 1, pp. 39–48. doi: 10.1016/j.apcata.2008.03.019.

Meher, L.C., D. Vidya Sagar, and S.N. Naik. 2006. "Technical Aspects of Biodiesel Production by Transesterification—A Review." *Renewable and Sustainable Energy Reviews* 10, no. 3, pp. 248–68. doi: 10.1016/j.rser.2004.09.002.

Mekhilef, S., S. Siga, and R. Saidur. 2011. "A Review on Palm Oil Biodiesel as a Source of Renewable Fuel." *Renewable and Sustainable Energy Reviews* 15, no. 4, pp. 1937–49. doi: 10.1016/j.rser.2010.12.012.

Melero, J.A., J. Iglesias, and G. Morales. 2009. "Heterogeneous Acid Catalysts for Biodiesel Production: Current Status and Future Challenges." *Green Chemistry* 11, no. 9, pp. 1285–1308. doi: 10.1039/B902086A.

Minami, E., and S. Saka. 2006. "Kinetics of Hydrolysis and Methyl Esterification for Biodiesel Production in Two-step Supercritical Methanol Process." *Fuel* 85, no. 17, pp. 2479–83. doi: 10.1016/j.fuel.2006.04.017.

Ming, L.O., H.M. Ghazali, and C. Chiew Let. 1999. "Use of Enzymatic Transesterified Palm Stearin-sunflower Oil Blends in the Preparation of Table Margarine Formulation." *Food Chemistry* 64, no. 1, pp. 83–88. doi: 10.1016/S0308-8146(98)00083-1.

Mittelbach, M., and P. Tritthart. 1988. "Diesel Fuel Derived from Vegetable Oils, III. Emission Tests Using Methyl Esters of Used Frying Oil." *Journal of the American Oil Chemists' Society* 65, no. 7, pp. 1185–7. doi: 10.1007/BF02660579.

Molitor, R., L. Clees, P. Czermak, and H. Koch. 2007. "Policy Instruments on Cars Energy Effiency.

Monteiro, M.R., A.R.P. Ambrozin, L.M. Lião, and A.G. Ferreira. 2008. "Critical Review on Analytical Methods for Biodiesel Characterization." *Talanta* 77, no. 2, pp. 593–605. doi: 10.1016/j.talanta.2008.07.001.

Mootabadi, H., B. Salamatinia, S. Bhatia, and A.Z. Abdullah. 2010. "Ultrasonic-assisted Biodiesel Production Process from Palm Oil Using Alkaline Earth Metal Oxides as the Heterogeneous Catalysts." *Fuel* 89, no. 8, pp. 1818–25.

Morgenstern, M., J. Cline, S. Meyer, and S. Cataldo. 2006. "Determination of the Kinetics of Biodiesel Production Using Proton Nuclear Magnetic Resonance Spectroscopy (1H NMR)." *Energy & Fuels* 20, no. 4, pp. 1350–3. doi: 10.1021/ef0503764.

Moser, B.R. 2011. "Biodiesel Production, Properties, and Feedstocks." In *Biofuels*, eds. D. Tomes, P. Lakshmanan, and D. Songstad, pp. 285–347. New York, NY: Springer.

Murphy, K.M. 2012. Analysis of Biodiesel Quality Using Reversed Phase High-Performance Liquid Chromatography.

Murphy, M.J., J.D. Taylor, and R.L. McCormick. 2004. *Compendium of Experimental Cetane Number Data*. Golden, CO: National Renewable Energy Laboratory, pp. 1–48.

Carvalho, M.S., M.A. Mendonça, D.M.M. Pinho, I.S. Resck, P.A.Z. Suarez. 2012. "Chromatographic Analyses of Fatty Acid Methyl Esters by HPLC-UV and

GC-FID." *Journal of Brazilian Chemistry Society* 23, no. 4. doi: 10.1590/S0103-50532012000400023.

Nabeel Adedapo, A., A.K.M. Mohiuddin, and A.T. Jameel. 2011. "Biodiesel Production: A Mini Review." *International Energy Journal (IEJ)* 12, no.1, pp. 15–28.

Narasimharao, K., A. Lee, and K. Wilson. 2007. "Catalysts in Production of Biodiesel: A Review." *Journal of Biobased Materials and Bioenergy* 1, no. 1, pp. 19–30. doi: 10.1166/jbmb.2007.002.

Nazir, N., N. Ramli, D. Mangunwidjaja, E. Hambali, D. Setyaningsih, S. Yuliani, and J. Salimon. 2009. "Extraction, Transesterification and Process Control in Biodiesel Production from Jatropha Curcas." *European Journal of Lipid Science and Technology* 111, no. 12, pp. 1185–200. doi: 10.1002/ejlt.200800259.

Ngamcharussrivichai, C., P. Totarat, and K. Bunyakiat. 2008. "Ca and Zn Mixed Oxide as a Heterogeneous Base Catalyst for Transesterification of Palm kernel Oil." *Applied Catalysis A: General* 341, no. 1, pp. 77–85. doi: 10.1016/j.apcata.2008.02.020.

Ngamprasertsith, S., and R. Sawangkeaw. 2011. "Transesterification in Supercritical Conditions." In *Biodiesel-Feedstocks and Processing Technologies*. Rijeka, Croatia: InTech, pp. 247–68.

Noiroj, K., P. Intarapong, A. Luengnaruemitchai, and S. Jai-In. 2009. "A Comparative Study of KOH/$A_{12}O_3$ and KOH/NaY Catalysts for Biodiesel Production via Transesterification from Palm Oil." *Renewable Energy* 34, no. 4, pp. 1145–50. doi: 10.1016/j.renene.2008.06.015.

Nollet, L.M., and F. Toldrá, (Eds.). 2012. *Food Analysis by HPLC* (Vol. 100). CRC Press.

Noureddini, H., and D. Zhu. 1997. "Kinetics of Transesterification of Soybean Oil." *Journal of the American Oil Chemists' Society* 74, no. 11, pp. 1457–63. doi: 10.1007/s11746-997-0254-2.

Obaja, D., S. Mace, J. Costa, C. Sans, and J. Mata-Alvarez. 2003. "Nitrification, Denitrification and Biological Phosphorus Removal in Piggery Wastewater Using a Sequencing Batch Reactor. *Bioresource Technology*, 87, no. 1, pp. 103–11. doi: 10.1016/S0960-8524(02)00229-8.

Odin, E.M., P.K. Onoja, and A.U. Ochala. 2013. "Effect of Process Variables on Biodiesel Production via Transesterification of Quassia Undulata Seed Oil, Using Homogeneous Catalyst." *International Journal of Scientific & Technology Research* 2, no. 9, pp. 267–76.

Oghome, P.I. 2012. "Transesterification and Optimization of Groundnut Oil Using Magnetic Stirrer." *Open Science Repository Engineering*, (open-access), e70081906.

Oh, P.P., H.L.N. Lau, J. Chen, M.F. Chong, and Y.M. Choo. 2012. "A Review on Conventional Technologies and Emerging Process Intensification (PI) Methods for Biodiesel Production." *Renewable and Sustainable Energy Reviews* 16, no. 7, pp. 5131–45. doi: 10.1016/j.rser.2012.05.014.

Ong, H.C., T.M.I. Mahlia, H.H. Masjuki, and R.S. Norhasyima. 2011. "Comparison of Palm Oil, Jatropha Curcas and Calophyllum Inophyllum for Biodiesel:

A Review." *Renewable and Sustainable Energy Reviews* 15, no. 8, pp. 3501–15. doi: 10.1016/j.rser.2011.05.005.

Pardal, A.C., J.M. Encinar, J.F. González, and G. Martínez. 2010. "Transesterification of Rapeseed Oil with Methanol in the Presence of Various Co-solvents." *Third International Symposium on Energy from Biomass and Waste.*Venice, Italy: CISA, Environmental Sanitary Engineering Centre..

Pascoli, S.D., A. Femia, and T. Luzzati. 2001. "Natural Gas, Cars and the Environment. A (Relatively)'Clean'and Cheap Fuel Looking for Users." *Ecological Economics* 38, no. 2, pp. 179–89. doi: 10.1016/S0921-8009(01)00174-4.

Peterson, C.L., J.L. Cook, J.C. Thompson, and J.S. Taberski. 2002. "Continuous Flow Biodiesel Production." *Applied Engineering in Agriculture* 18, no. 1, pp. 5–11. doi: 10.13031/2013.7702.

Peterson, C.L., D.L. Reece, B.L. Hammond, J. Thompson, and S.M. Beck. 1997. "Processing, Characterization, and Performance of Eight Fuels from Lipids." *Applied Engineering in Agriculture* 13, no. 1, pp. 71–79. doi: 10.13031/2013.21578.

Petrou, E.C., and C.P. Pappis. 2009. "Biofuels: A Survey on Pros and Cons." *Energy & Fuels* 23, no. 2, pp. 1055–66. doi: 10.1021/ef800806g.

Phan, A.N., A.P. Harvey, and M. Rawcliffe. 2011. "Continuous Screening of Base-catalysed Biodiesel Production Using New Designs of Mesoscale Oscillatory Baffled Reactors." *Fuel Processing Technology* 92, no. 8, pp. 1560–7. doi: 10.1016/j.fuproc.2011.03.022.

Pinto, A.C., L.L. Guarieiro, M.J. Rezende, N.M. Ribeiro, E.A. Torres, W.A. Lopes, and J.B.D. Andrade. 2005. "Biodiesel: An Overview." *Journal of the Brazilian Chemical Society* 16, no. 6B, pp. 1313–30. doi: 10.1590/S0103-50532005000800003.

Poljanšek, I., and B. Likozar. 2011. "Influence of Mass Transfer and Kinetics on Biodiesel Production Process." In *Mass Transfer in Multiphase Systems and its Applications*, pp. 433–58. Croatia: Intech.

Portnoff, M.A., D.A. Purta, M.A. Nasta, J. Zhang, and F. Pourarian. 2005. Methods for Producing biodiesel. US Patent 20,050,274,065. Washington, DC: U.S. Patent and Trademark Office.

Predojević, Z.J. 2008. "The Production of Biodiesel from Waste Frying Oils: A Comparison of Different Purification Steps." *Fuel* 87, no. 17, pp. 3522–8. doi: 10.1016/j.fuel.2008.07.003.

Pryde, E.H. 1983. "Vegetable Oils as Diesel Fuels: Overview." *Journal of the American Oil Chemists' Society* 60, no. 8, pp. 1557–8. doi: 10.1007/BF02666584.

Qiu, Z., J. Petera, and L.R. Weatherley. 2012. "Biodiesel Synthesis in an Intensified Spinning Disk Reactor." *Chemical Engineering Journal* 210, pp. 597–609. doi: 10.1016/j.cej.2012.08.058.

Ragonese, C., P.Q. Tranchida, D. Sciarrone, and L. Mondello. 2009. "Conventional and Fast Gas Chromatography Analysis of Biodiesel Blends Using an Ionic Liquid Stationary Phase." *Journal of Chromatography A*, 1216, no. 51, pp. 8992–7. doi: 10.1016/j.chroma.2009.10.066.

Rashid, U., and F. Anwar. 2008. "Production of Biodiesel through Base-catalyzed Transesterification of safflower oil Using an Optimized Protocol." *Energy & Fuels* 22, no. 2, pp. 1306–12. doi: 10.1021/ef700548s.

Reay, D. 2008. "The Role of Process Intensification in Cutting Greenhouse Gas Emissions." *Applied Thermal Engineering* 28, no. 16, pp. 2011–9. doi: 10.1016/j.applthermaleng.2008.01.004.

Refaat, A.A., S.T. El Sheltawy, and K.U. Sadek. 2008. "Optimum Reaction Time, Performance and Exhaust Emissions of Biodiesel Produced by Microwave Irradiation." *International Journal of Environmental Science & Technology* 5, no. 3, pp. 315–22. doi: 10.1007/BF03326026.

Rezanka, T., and K. Sigler. 2007. "The Use of Atmospheric Pressure Chemical Ionization Mass Spectrometry with High Performance Liquid Chromatography and Other Separation Techniques for Identification of Triacylglycerols." *Current Analytical Chemistry* 3, no. 4, pp. 252–71. doi: 10.2174/157341107782109644.

Royon, D., M. Daz, G. Ellenrieder, and S. Locatelli. 2007. "Enzymatic Production of Biodiesel from Cotton Seed Oil Using Butanol as a Solvent." *Bioresource Technology* 98, no. 3, pp. 648–53. doi: 10.1016/j.biortech.2006.02.021.

Russell, T.W. 1945. Process of Treating Fatty Glycerides. US Patent 2,383,632. Washington, DC: U.S. Patent and Trademark Office.

Saka, S., and D. Kusdiana. 2001. "Biodiesel Fuel from Rapeseed Oil as Prepared in Supercritical Methanol." *Fuel* 80, no. 2, pp. 225–31. doi: 10.1016/S0016-2361(00)00083-1.

Salamatinia, B., H. Mootabadi, S. Bhatia, and A.Z. Abdullah. 2010. "Optimization of Ultrasonic-assisted Heterogeneous Biodiesel Production from Palm Oil: A Response Surface Methodology Approach." *Fuel Processing Technology* 91, no. 5, pp. 441–8. doi: 10.1016/j.fuproc.2009.12.002.

Šalić, A., and B. Zelić. 2011. "Microreactors-portable Factories for Biodiesel Fuel Production." *goriva i maziva* 50, no. 2, pp. 98–110.

Salzano, E., M. Di Serio, and E. Santacesaria. 2010. "Emerging Risks in the Biodiesel Production by Transesterification of Virgin and Renewable Oils." *Energy & Fuels* 24, no. 11, pp. 6103–9. doi: 10.1021/ef101229b.

Samart, C., P. Sreetongkittikul, and C. Sookman. 2009. "Heterogeneous Catalysis of Transesterification of Soybean Oil Using KI/mesoporous Silica." *Fuel Processing Technology* 90, no. 7, pp. 922–5. doi: 10.1016/j.fuproc.2009.03.017.

Sarin, A., R. Arora, N.P. Singh, M. Sharma, and R.K. Malhotra. 2009. "Influence of Metal Contaminants on Oxidation Stability of Jatropha Biodiesel." *Energy* 34, no. 9, pp. 1271–5. doi: 10.1016/j.energy.2009.05.018.

Savage, P.E., S. Gopalan, T.I. Mizan, C.J. Martino, and E.E. Brock. 1995. "Reactions at Supercritical Conditions: Applications and Fundamentals." *AIChE Journal* 41, no. 7, pp. 1723–78. doi: 10.1002/aic.690410712.

Scholfield, C.R. 1975. "High Performance Liquid Chromatography of Fatty Methyl Esters: Analytical Separations." *Journal of the American Oil Chemists' Society* 52, no. 2, pp. 36–37. doi: 10.1007/BF02901818.

Schuchardt, U., R. Sercheli, and R.M. Vargas. 1998. "Transesterification of Vegetable Oils: A Review." *Journal of the Brazilian Chemical Society* 9, no. 3, pp. 199–210. doi: 10.1590/S0103-50531998000300002.

Seeley, J.V., S.K. Seeley, E.K. Libby, and J.D. McCurry. 2007. "Analysis of Biodiesel/Petroleum Diesel Blends with Comprehensive Two-dimensional Gas Chromatography." *Journal of chromatographic Science* 45, no. 10, pp. 650–6. doi: 10.1093/chromsci/45.10.650.

Senatore, A., M. Cardone, V. Rocco, and M.V. Prati. 2000. A Comparative Analysis of Combustion Process in DI Diesel Engine Fueled with Biodiesel and Diesel Fuel (No. 2000-01-0691). SAE Technical Paper.

Şensöz, S., D. Angın, and S. Yorgun. 2000. "Influence of Particle Size on the Pyrolysis of Rapeseed Brassica Napus: Fuel Properties of Bio-oil." *Biomass and Bioenergy* 19, no. 4, pp. 271–9. doi: 10.1016/S0961-9534(00)00041-6.

Serrano, M., A. Bouaid, M. Martínez, and J. Aracil. 2013. "Oxidation Stability of Biodiesel from Different Feedstocks: Influence of Commercial Additives and Purification Step." *Fuel* 113, pp. 50–58. doi: 10.1016/j.fuel.2013.05.078.

Shah, S., and M.N. Gupta. 2007. "Lipase Catalyzed Preparation of Biodiesel from Jatropha Oil in a Solvent Free System." *Process Biochemistry* 42, no. 3, pp. 409–14. doi: 10.1016/j.procbio.2006.09.024.

Sharma, B.K., B.R. Moser, K.E. Vermillion, K.M. Doll, and N. Rajagopalan. 2014. "Production, Characterization and Fuel Properties of Alternative Diesel Fuel from Pyrolysis of Waste Plastic Grocery Bags." *Fuel Processing Technology* 122, pp. 79–90. doi: 10.1016/j.fuproc.2014.01.019.

Sharma, Y.C., B. Singh, and S.N. Upadhyay. 2008. "Advancements in Development and Characterization of Biodiesel: A Review." *Fuel* 87, no. 12, pp. 2355–73. doi: 10.1016/j.fuel.2008.01.014.

Sheikh, R., M.S. Choi, J.S. Im, and Y.H. Park. 2013. "Study on the Solid Acid Catalysts in Biodiesel Production from High Acid Value Oil." *Journal of Industrial and Engineering Chemistry* 19, no. 4, pp. 1413–9. doi: 10.1016/j.jiec.2013.01.005.

Shin, H.Y., S.H. Lee, J.H. Ryu, and S.Y. Bae. 2012. "Biodiesel Production from Waste Lard Using Supercritical Methanol." *The Journal of Supercritical Fluids* 61, pp. 134–38. doi: 10.1016/j.supflu.2011.09.009.

Shuit, S.H., Y.T. Ong, K.T. Lee, B. Subhash, and S.H. Tan. 2012. "Membrane Technology as a Promising Alternative in Biodiesel Production: A review." *Biotechnology Advances* 30, no. 6, 1364–80. doi: 10.1016/j.biotechadv.2012.02.009.

Singh, A.K., S.D. Fernando, and R. Hernandez. 2007. "Base-catalyzed Fast Transesterification of Soybean Oil Using Ultrasonication." *Energy & Fuels* 21, no. 2, pp. 1161–4. doi: 10.1031/2013.21551.

Singh, A., B. He, J. Thompson, and J. Van Gerpen. 2006. "Process Optimization of Biodiesel Production Using Alkaline Catalysts." *Applied Engineering in Agriculture* 22, no. 4, pp. 597–600. doi: 10.13031/2013.21213.

Singh, A., P.S. Nigam, and J.D. Murphy. 2011. "Renewable Fuels from Algae: An Answer to Debatable Land Based Fuels." *Bioresource Technology* 102, no. 1, pp. 10–16. doi: 10.1016/j.biortech.2010.06.032.

Sivasamy, A., K.Y. Cheah, P. Fornasiero, F. Kemausuor, S. Zinoviev, and S. Miertus. 2009. "Catalytic Applications in the Production of Biodiesel from Vegetable Oils." *ChemSusChem* 2, no. 4, pp. 278–300. doi: 10.1002/cssc.200800253.

Snyder, D.B., G.H. Adi, C.M. Hall, M.P. Bunce, and G.M. Shaver. January 2010. "Closed-Loop Control Framework for Fuel-Flexible Combustion of Biodiesel Blends." In *ASME 2010 Internal Combustion Engine Division Fall Technical Conference* (pp. 665–674). American Society of Mechanical Engineers.

Soriano, N.U. Jr, R. Venditti, and D.S. Argyropoulos. 2009. "Biodiesel Synthesis via Homogeneous Lewis Acid-catalyzed Transesterification." *Fuel* 88, no. 3, pp. 560–5. doi: 10.1016/j.fuel.2008.10.013.

Stamenković, O.S., M.L. Lazić, Z.B. Todorović, V.B. Veljković, and D.U. Skala. 2007. "The Effect of Agitation Intensity on Alkali-catalyzed Methanolysis of Sunflower Oil." *Bioresource Technology* 98, no. 14, pp. 2688–99. doi: 10.1016/j.biortech.2006.09.024.

Stern, R., and G. Hillion. 1990. Purification of Esters: Eur Pat Appl EP 356317. *Cl. C07C67/56) [cf Chem Abstr 113: 58504k]*.

Stojković, I.J., O.S. Stamenković, D.S. Povrenović, and V.B. Veljković. 2014. "Purification Technologies for Crude Biodiesel Obtained by Alkali-catalyzed Transesterification." *Renewable and Sustainable Energy Reviews* 32, pp. 1–15. doi: 10.1016/j.rser.2014.01.005.

Stoytcheva, M. 2011. *Biodiesel–Quality, Emissions and By-products*. Croatia: Intech.

Sun, J., J. Ju, L. Ji, L. Zhang, and N. Xu. 2008. "Synthesis of Biodiesel in Capillary Microreactors". *Industrial & Engineering Chemistry Research* 47, no. 5, pp. 1398–403. doi: 10.1021/ie070295q.

Taft, R.W. Jr, M.S. Newman, and F.H. Verhoek. 1950. "The Kinetics of the Base-catalyzed Methanolysis of Ortho, Meta and Para Substituted l-Menthyl Benzoates1, 2." *Journal of the American Chemical Society* 72, no. 10, pp. 4511–19.

Tamalampudi, S., M.R. Talukder, S. Hama, T. Numata, A. Kondo, and H. Fukuda. 2008. "Enzymatic Production of Biodiesel from Jatropha oil: A Comparative Study of Immobilized-whole Cell and Commercial Lipases as a Biocatalyst." *Biochemical Engineering Journal* 39, no. 1, pp. 185–9.

Tan, H.W., A.R. Abdul Aziz, and M.K. Aroua. 2013. "Glycerol Production and its Applications as a Raw Material: A Review." *Renewable and Sustainable Energy Reviews* 27, pp. 118–27. doi: 10.1016/j.rser.2013.06.035.

Teixeira, L.S., J.C. Assis, D.R. Mendonça, I.T. Santos, P.R. Guimarães, L.A. Pontes, and J.S. Teixeira. 2009. "Comparison Between Conventional and Ultrasonic Preparation of Beef Tallow Biodiesel." *Fuel Processing Technology* 90, no. 9, pp. 1164–6. doi: 10.1016/j.fuproc.2009.05.008.

Ter Horst, M., S. Urbin, R. Burton, and C. McMillan. 2009. "Using Proton Nuclear Magnetic Resonance as a Rapid Response Research Tool for Methyl Ester Characterization in Biodiesel." *Lipid Technology* 21, no. 2, pp. 39–41. doi: 10.1002/lite.200900004.

Terry, B., R.L. McCormick, and M. Natarajan. 2006. Impact of Biodiesel Blends on Fuel System Component Durability (No. 2006-01-3279). SAE Technical Paper.

Thompson, J.C., and B.B. He. 2007. "Biodiesel Production Using Static Mixers." *Transactions of the ASABE* 50, no. 1, pp. 161–5. doi: 10.13031/2013.22389.

Tilman, A., H. Dietz, M. Hahl, N. Nikolidakis, C. Rosendah, and K. Seelige. 2009. *Biodiesel In India: Value Chain Organisation and Policy Options for Rural Development.* German Development Institute.

Tomasevic, A.V., and S.S. Siler-Marinkovic. 2003. "Methanolysis of Used Frying Oil." *Fuel Processing Technology* 81, no. 1, pp. 1–6. doi: 10.1016/S0378-3820(02)00096-6.

Trevisan, M.G., C.M. Garcia, U. Schuchardt, and R.J. Poppi. 2008. "Evolving Factor Analysis-Based Method for Correcting Monitoring Delay in Different Batch Runs for Use With PLS: On-Line Monitoring of a Transesterification Reaction by ATR-FTIR." *Talanta* 74, no. 4, pp. 971–6.

Turck, R. 2003. Method for Producing Fatty Acid Esters of Monovalent Alkyl Alcohols and Use thereof. US Patent 6,538,146. Washington, DC: U.S. Patent and Trademark Office.

Van de Velde, L., W. Verbeke, M. Popp, J. Buysse, and G. Van Huylenbroeck. 2009. "Perceived Importance of Fuel Characteristics and Its Match with Consumer Beliefs About Biofuels in Belgium." *Energy Policy* 37, no. 8, pp. 3183–93.

Van Gerpen, J. 2005. Biodiesel Production and Fuel Quality. *Department of Biological and Agricultural Engineering University of Idaho, Moscow, ID.*

Van Gerpen, J.H., and B. He. 2010. "Biodiesel Production and Properties." In *Thermochemical Conversion of Biomass to Liquid Fuels and Chemicals*, pp. 382–415. Cambridge, Canada: RSC Publishing.

Van Gerpen, J.H., E.G. Hammond, L. Yu, and A. Monyem. 1997. Determining The Influence of Contaminants on Biodiesel Properties (No. 971685). SAE Technical Paper.

Van Gerpen, J.H., R. Pruszko, D. Clements, B. Shanks, and G. Knothe. 2006. *Building a Successful Biodiesel Business.* Biodiesel Basics.

Van Gerpen, J., and G. Knothe. 2005. "Basics of the transesterification reaction." In *The Biodiesel Handbook*, eds. G. Knothe, J. Van Gerpen, and J. Krahl, pp. 26–41. Champaign, IL: AOCS Press.

Van Kasteren, J.M.N., and Nisworo, A.P. 2007. "A Process Model to Estimate the Cost of Industrial Scale Biodiesel Production from Waste Cooking Oil by Supercritical Transesterification." *Resources, Conservation and Recycling* 50, no. 4, pp. 442–58. doi: 10.1016/j.resconrec.2006.07.005.

Van Walwijk, M. 2005. Biofuels in France. Subcontract Report for Premia, France.

Veljković, V.B., J.M. Avramović, and O.S. Stamenković. 2012. "Biodiesel Production by Ultrasound-assisted Transesterification: State of the Art and the Perspectives." *Renewable and Sustainable Energy Reviews* 16, no. 2, pp. 1193–209. doi: 10.1016/j.rser.2011.11.022.

Vicente, G., M. Martınez, and J. Aracil. 2004. "Integrated Biodiesel Production: A Comparison of Different Homogeneous Catalysts Systems." *Bioresource Technology* 92, no. 3, pp. 297–305. doi: 10.1016/j.biortech.2003.08.014.

Vyas, A.P., J.L. Verma, and N. Subrahmanyam. 2010. "A Review on FAME Production Processes." *Fuel* 89, no. 1, pp. 1–9. doi: 10.1016/j.fuel.2009.08.014.

Walker, D.A. (2009). Biofuels, facts, fantasy, and feasibility. *Journal of Applied Phycology* 21(5), 509–17. doi: 10.1016/j.fuel.2009.08.014.

Wang, Y., X. Wang, Y. Liu, S. Ou, Y. Tan, and S. Tang. 2009. "Refining of Biodiesel by Ceramic Membrane Separation." *Fuel Processing Technology* 90, no. 3, pp. 422–7.doi: 10.1016/j.fuproc.2008.11.004.

Wen, Z., X. Yu, S.T. Tu, J. Yan, and E. Dahlquist. 2009. "Intensification of Biodiesel Synthesis Using Zigzag Micro-channel Reactors." *Bioresource Technology* 100, no. 12, pp. 3054–60. doi 10.1016/j.biortech.2009.01.022.

Westbrook, S.R. 2005. Evaluation and Comparison of Test Methods to Measure the Oxidation Stability of Neat Biodiesel. United States. Department of Energy.

Wimmer, T. 1995. Process for the Production of Fatty Acid Esters of Lower Alcohols. US Patent 5,399,731. Washington, DC: U.S. Patent and Trademark Office.

*www.extension.org/pages/.../used-and-waste-oil-and-grease-for-biodiesel*May 28, 2014.

NextCAT-Making Biodiesel Profitable, http://www.nextcatinc.com/The%20Next-CAT%20Solution.htm

XiaoHu, F., W. Xi, and C. Feng. 2011. "Biodiesel Production from Crude Cottonseed Oil: An Optimization Process Using Response Surface Methodology." *The Open Fuels & Energy Science Journal* 4, pp. 1–8. doi: 10.2174/1876973X01104010001.

Xie, W., and H. Li. 2006. "Alumina-supported Potassium Iodide as a Heterogeneous Catalyst for Biodiesel Production from Soybean Oil." *Journal of Molecular Catalysis A: Chemical* 255, no. 1, pp. 1–9. doi: 10.1016/j.molcata.2006.03.061.

Xie, W., H. Peng, and L. Chen. 2006. "Calcined Mg–Al Hydrotalcites as Solid Base Catalysts for Methanolysis of Soybean Oil." *Journal of Molecular Catalysis A: Chemical* 246, no. 1, pp. 24–32. doi: 10.1016/j.molcata.2005.10.008.

Xie, W., Z. Yang, and H. Chun. 2007. "Catalytic Properties of Lithium-doped ZnO Catalysts Used for Biodiesel Preparations." *Industrial & Engineering Chemistry Research* 46, no. 24, 7942–9. doi: 10.1021/ie070597s.

Xing-cai, L., Y. Jian-Guang, Z. Wu-Gao, and H. Zhen. 2004. "Effect of Cetane Number Improver on Heat Release Rate and Emissions of High Speed Diesel Engine Fueled with Ethanol–Diesel Blend Fuel." *Fuel* 83, no. 14, pp. 2013–20. doi: 10.1016/j.fuel.2004.05.003.

Xu, L., Y. Wang, X. Yang, J. Hu, W. Li, and Y. Guo. 2009. "Simultaneous Esterification and Transesterification of Soybean Oil with Methanol Catalyzed by Mesoporous Ta2O5/SiO2–[H3PW12O40](R= Me or Ph) Hybrid Catalysts." *Green Chemistry* 11, no. 3, pp. 314–7. doi: 10.1039/B815279A.

Yadav, G.D., and A.D. Murkute. 2004. "Preparation of a Novel Catalyst UDCaT-5: Enhancement in Activity of Acid-treated Zirconia—Effect of Treatment with Chlorosulfonic Acid vis-à-vis Sulfuric acid." *Journal of Catalysis* 224, no. 1, pp. 218–23. doi: 10.1016/j.jcat.2004.02.021.

Yamane, K., A. Ueta, and Y. Shimamoto. 2001. "Influence of Physical and Chemical Properties of Biodiesel Fuels on Injection, Combustin and Exhaust Emission Characteristics in a Direct Injection Compression Ignition

Engine." *International Journal of Engine Research* 2, no. 4, pp. 249–61. doi: 10.1243/1468087011545460.

Yee, K.F., K.T. Tan, A.Z. Abdullah, and K.T. Lee. 2009. "Life Cycle Assessment of Palm Biodiesel: Revealing Facts and Benefits for Sustainability." *Applied Energy* 86, pp. S189–S196. doi: 10.1016/j.apenergy.2009.04.01.

Yori, J.C., S.A. D'ippolito, C.L. Pieck, and C.R. Vera. 2007. "Deglycerolization of Biodiesel Streams by Adsorption over Silica Beds." *Energy & Fuels* 21, no. 1, pp. 347–53. doi: 10.1021/ef060362d.

Yu, D., L. Tian, H. Wu, S. Wang, Y. Wang, D. Ma, and X. Fang. 2010. "Ultrasonic Irradiation with Vibration for Biodiesel Production from Soybean Oil by Novozym 435. *Process Biochemistry* 45, no. 4, pp. 519–25. doi: 10.1016/j.procbio.2009.11.012.

Yuan, H., B.L. Yang, and G.L. Zhu. 2008. "Synthesis of Biodiesel Using Microwave Absorption Catalysts." *Energy & Fuels* 23, no. 1, pp. 548–52. doi: 10.1021/ef800577j.

Zabaniotou, A., O. Ioannidou, and V. Skoulou. 2008. "Rapeseed Residues Utilization for Energy and 2nd Generation Biofuels." *Fuel* 87, no. 8, pp. 1492–502. doi: 10.1016/j.fuel.2007.09.003

Zawadzki, A., D.S. Shrestha, and B. He. 2007. "Biodiesel Blend Level Detection Using Ultraviolet Absorption Spectra." *Transactions of the ASABE* 50, no. 4, pp. 1349–53. doi: 10.13031/2013.23612.

Zeng, W., M. Xu, M. Zhang, Y. Zhang, and D. Cleary. 2010. Characterization of Methanol and Ethanol Sprays from Different Di Injectors by Using Mie-scattering and Laser Induced Fluorescence at Potential Engine Cold-start Conditions (no. 2010-01-0602). SAE Technical Paper.

Zhang, Y., M.A. Dube, D.D. McLean, and M. Kates. 2003a. "Biodiesel Production from Waste Cooking Oil: 2. Economic Assessment and Sensitivity Analysis." *Bioresource Technology* 90, no. 3, pp. 229–40. doi: 10.1016/S0960-8524(03)00150-0.

Zhang, Y., M.A. Dube, D. McLean, and M. Kates. 2003b. "Biodiesel Production from Waste Cooking Oil: 1. Process Design and Technological Assessment." *Bioresource Technology* 89, no. 1, pp. 1–16. doi: 10.1016/S0960-8524(03)00040-3.

Zhang, Y., M. Stanciulescu, and M. Ikura. 2009. "Rapid Transesterification of Soybean Oil with Phase Transfer Catalysts." *Applied Catalysis A: General* 366, no. 1, pp. 176–83. doi: 10.1016/j.apcata.2009.07.001.

Zheng, Y., X.M. Wu, C. Branford-White, J. Quan, and L.M. Zhu. 2009. "Dual Response Surface-Optimized Process for Feruloylated Diacylglycerols by Selective Lipase-catalyzed Transesterification in Solvent Free System." *Bioresource Technology* 100, no. 12, pp. 2896–901.

Zhou, W., S.K. Konar, and D.G. Boocock. 2003. "Ethyl Esters from the Single-phase Base-catalyzed Ethanolysis of Vegetable Oils." *Journal of the American Oil Chemists' Society* 80, no. 4, pp. 367–71.

ABOUT THE AUTHORS

Aminul Islam holds an MPhil in Materials Science from the Bangladesh University of Engineering and Technology, Bangladesh, in 2008. He received his PhD in from the Universiti Malaysia Sabah, Malaysia, in 2012. Since 2012, he has been a postdoctoral fellow researcher at the Catalysis Science and Technology Research Centre, Universiti Putra Malaysia (Malaysia). His scientific interests are focused on biodiesel, renewable energy, technological applications of materials science, and environmental heterogeneous catalysis.

Yun Hin Taufiq-Yap is professor of catalysis and head of the Catalysis Science and Technology Research Centre at Universiti Putra Malaysia. He is a Fellow of Malaysia Institute of Chemistry and Royal Society of Chemistry, UK. He holds a BSc (Hons) and MSc from Universiti Putra Malaysia and a PhD from the University of Manchester Institute of Science and Technology (UMIST), UK and following research attachment at Cardiff University. He is currently a visiting professor for Curtin Sarawak Research Institute at Curtin University Sarawak and was also formerly a visiting professor at Nagoya University, Japan and at Universiti Teknologi PETRONAS. His research interests lie on designing heterogeneous catalysts and nanocatalyst for sustainable biofuels and chemicals production from biomass and renewable resources.

Eng-Seng Chan received his BEng (1998) and PhD (2002) from the University of Birmingham (UK). He is an associate professor and the head of discipline (Chemical Engineering) at Monash University Malaysia. His research interest revolves around the development of particulate systems for applications in biofuel production, food processing, water purification, protein purification, and drug and food delivery.

INDEX

FORTHCOMING TITLES FROM OUR THERMAL SCIENCE AND ENERGY ENGINEERING COLLECTION

Derek Dunn-Rankin, Editor

Advanced Technologies in Biodiesel: New Advances
in Designed and Optimized Catalysts
By Aminul Islam

Optimization of Cooling Systems
By David Zietlow

Graphical Thermodynamics
By Moufid Hilal

Momentum Press publishes several other collections, including:
Industrial, Systems, and Innovation Engineering; Manufacturing and
Processes; Engineering Management; Electrical Power; Fluid Mechanics;
Acoustical Engineering; Aerospace Engineering; Biomedical Engineering;
and Healthcare Administration.

*Momentum Press is actively seeking collection editors as well as authors. For
more information about becoming an MP author or collection editor, please
visit http://www.momentumpress.net/contact*

Announcing Digital Content Crafted by Librarians

Momentum Press offers digital content as authoritative treatments of advanced engineering topics by leaders in their field. Hosted on ebrary, MP provides practitioners, researchers, faculty, and students in engineering, science, and industry with innovative electronic content in sensors and controls engineering, advanced energy engineering, manufacturing, and materials science.

Momentum Press offers library-friendly terms:

- perpetual access for a one-time fee
- no subscriptions or access fees required
- unlimited concurrent usage permitted
- downloadable PDFs provided
- free MARC records included
- free trials

The **Momentum Press** digital library is very affordable, with no obligation to buy in future years.

For more information, please visit **www.momentumpress.net/library** or to set up a trial in the US, please contact **mpsales@globalpress.com**.